WHY DID IT HAVE TO BE SNAKES?

Books by Lois H. Gresh and Robert Weinberg

The Science of Stephen King
The Science of James Bond
The Science of Superheroes
The Science of Supervillains
The Science of Anime
The Termination Node
The Computers of Star Trek

Books by Lois H. Gresh

Chuck Farris and the Tower of Darkness
Chuck Farris and the Labyrinth of Doom
Chuck Farris and the Cosmic Storm
The Truth behind a Series of Unfortunate Events
Dragonball Z
TechnoLife 2020
The Ultimate Unauthorized Eragon Guide
The Fan's Guide to the Spiderwick Chronicles
Exploring Philip Pullman's His Dark Materials

Books by Robert Weinberg

Secrets of X-Men Revealed
A Logical Magician
A Calculated Magic
The Black Lodge
The Dead Man's Kiss
The Devil's Auction
The Armageddon Box

WHY DID IT HAVE TO BE SNAKES?

From SCIENCE to the SUPERNATURAL, the MANY MYSTERIES of INDIANA JONES

Lois H. Gresh
and
Robert Weinberg

WILEY

John Wiley & Sons, Inc.

Published by John Wiley & Sons, Inc., Hoboken, New Jersey

Published simultaneously in Canada

For general information about our other products and services, please contact our Customer Care Department within the United States at (800) 762-2974, outside the United States at (317) 572-3993 or fax (317) 572-4002.

Wiley also publishes its books in a variety of electronic formats. Some content that appears in print may not be available in electronic books. For more information about Wiley products, visit our web site at www.wiley.com.

Library of Congress Cataloging-in-Publication Data:

Gresh, Lois H.
   Why did it have to be snakes? : from science to the supernatural, the many mysteries of Indiana Jones / Lois Gresh and Robert Weinberg.
      p. cm.
   Includes bibliographical references and index.
   ISBN 978-1-62045-563-0
1. Indiana Jones films—Miscellanea.   I. Weinberg, Robert E.   II. Title.
PN1995.9.I47G74 2008
791.45'75—dc22   2007039346

Printed in the United States of America

10 9 8 7 6 5 4 3 2

Contents

Insects ◇ Thuggees and Kali ◇ Ripping a Heart out of a Live Body ◇ Drinking Blood ◇ Zombies ◇ Voodoo Dolls ◇ Crocodiles

Acknowledgments

We would like to thank Arie Bodek for specifics about the Urim and Thummim, derived directly from his copy of the Torah; and Kevin L. O'Brien and Angeline Hawkes for their assistance in researching the Great Seal. Special thanks to Dan Gresh for his patience over the years while his mother wrote books on weekends and at night.

Thanks as always to our agent, Lori Perkins, and to Stephen S. Power, our editor at John Wiley & Sons.

INTRODUCTION

Everyone knows that Harrison Ford plays the role of Indiana Jones, an adventurer and an archaeologist whose bravery has no limits. Indiana sports a leather jacket and a fedora, he brandishes a whip, and he can do most anything. His biggest fear in life is snakes. Millions of people all over the world were introduced to the character in 1981, when Paramount Pictures released *Raiders of the Lost Ark*, later retitled for video *Indiana Jones and the Raiders of the Lost Ark*. The film was directed by Steven Spielberg, the story was developed by George Lucas and Philip Kaufman, and the official screenwriter was Lawrence Kasdan. With a duo like Spielberg and Lucas at the helm, it was no surprise that Indiana Jones became an icon in the public mind.

The movie was nominated for eight Academy Awards, including Best Picture, and it won Oscars for Best Sound, Best Film Editing, Best Visual Effects, and Best Art Direction-Set Decoration. In addition, the film won a Special Achievement Award for Sound Effects Editing.

As the story goes, Henry "Indiana" Jones Jr. was born in Princeton, New Jersey. His mother was Anna Jones, who died when Indy was a child, and his father was Henry Jones. Indy's original screen name was going to be Indiana Smith, after George Lucas's Alaskan malamute dog, Indiana. Indiana Jones is of average build (that is, not huge), is incredibly strong, is extremely intelligent, and needs a lot of physical and mental stimulation. And he's not exactly a conformist, either.

Indiana's love interest in *Raiders of the Lost Ark* was Marion Ravenwood, played by the actress Karen Allen. The Nazi-commissioned archaeologist Dr. Rene Belloq was played by Paul

Freeman. Other actors in the film included Ronald Lacey as Major Toht and John Rhys-Davies as Sallah. The film dealt with Nazis trying to locate the lost biblical Ark of the Covenant and Indy's attempts to make sure they didn't find it. The adventure took place in the mountains of Tibet, the desert in Egypt, and a secret Nazi submarine base in the Mediterranean Sea. The story was fast-paced and the action nonstop. The film was a huge hit.

Indiana Jones's second adventure was a prequel called *Indiana Jones and the Temple of Doom*. The story was once again conceived by Lucas and directed by Spielberg. The film also included Kate Capshaw as the ever-terrified, ever-screaming Willie Scott and Jonathan Ke Quan as Indiana's young helper, Short Round. The film, released by Paramount in 1984, won an Academy Award for Visual Effects. Although it cost $8 million more to make than *Raiders* did, a now seemingly paltry $28 million, it received mixed reviews, and the U.S. gross was $65 million less, albeit a still substantial $180 million.

Temple of Doom was so dark that it was supposedly almost called *Temple of Death*. It included child slavery, torture, and gory scenes such as the one in which a Thuggee high priest yanks a beating, bloody heart from a victim's chest. Because the Motion Picture Association of America (MPAA) had only PG and R ratings at the time, Spielberg asked the MPAA to make a new rating, which became PG-13.

Temple of Doom is an excellent example of the outdated Western view of Asian society. In the Fu Manchu series, for example, the typical Western view was that Asian people eat weird exotic foods, are members of gangs, and kill with no remorse or feelings. In *Temple of Doom*, Indians are cast in these stereotypical shades, to the point where the country of India banned the film. Eventually, the ban was lifted.

The third film in the series, *Indiana Jones and the Last Crusade*, was released in 1989 and set in 1938. In this film, Indiana Jones must find the Holy Grail with the help of his father, played by Sean Connery, and with no thanks to Dr. Marcus Brody, a friend of both Indy and his father, played by Denholm Elliott. The movie opens with a younger Indiana Jones, played by River Phoenix, acquiring his trademark bullwhip, fedora, and fear of snakes. It was the first film in history to gross $50 million in one week, and it went on to gross $197 million in the United States.

May 22, 2008, will see the release of the fourth movie, titled *Indiana Jones and the Kingdom of the Crystal Skull*. The movie, set in 1957, will reunite Harrison Ford with his former costar Karen Allen, who will reprise her role as Marion Ravenwood.

The television series *The Young Indiana Jones Chronicles* ran from 1992 to 1996 with the ten-year-old Jones played by Corey Carrier and the seventeen-year-old Jones by Sean Patrick Flanery. The forty-four-episode series consisted of twenty-two stories, each two parts long. In these stories, Indy met a great number of historical figures and participated in many of the important events of the early twentieth century. The series was released on DVD in 2007.

Surprisingly, the DVD set of the first three Indiana Jones movies includes no special features about the historical, cultural, or scientific backgrounds of the films. A fourth DVD in the set contains three hours of material about how the films were produced, how the special effects were shot, and even how the music was scored.

When it comes to how the script was written, however, there is nothing on the background material of the movies. The only reference to content is that Spielberg and Lucas wanted to make the movies in the style of the 1930s serials. Not a word is spoken about keeping the films accurate to the times or researching the background of the iconic images and settings that appear throughout all three films.

Numerous questions are raised by watching the movies, and no answers are provided. Unless viewers are well versed in the history, the culture, and the science of the times, most of the references and the background of the movies make little sense.

This book is the first publication to answer the numerous questions that have troubled Indiana Jones fans for years. We tell you about the fedora, the bullwhip, and the snakes. We discuss the Thuggees and their deadly habits, ancient death traps, the Nazis and the occult, the Well of Souls, the Ark of the Covenant, gangs in 1935 Shanghai, Kali worship in India, the infamous, bizarre banquet in *Temple of Doom* that included chilled monkey brains, the

Sankara Stones, the Cross of Coronado, the Last Supper, the Holy Grail, and much more. Anything you want to know about the science and the history behind the Indiana Jones saga is in this book, so turn the pages and begin to explore the adventures and many mysteries of Indiana Jones.

PART 1

INDIANA JONES

and the
Raiders of the Lost Ark

The Hovitos Temple and the Golden Idol

The year is 1936, and Professor Indiana Jones is braving the dangers of the Peruvian jungle to retrieve the precious golden Hovitos idol from an ancient temple.

With some Indian workmen and guides Satipo and Barranca, Indy makes his way through the dense underbrush of the jungle. As one of the Indians parts some branches, exposing a gruesome-looking statue, birds scatter and the Indian screams in terror. Indy then finds a dart stuck in a tree, and Satipo identifies the poison on the dart as "fresh, three days." Soon, the group discovers a cave entrance to a temple, and Satipo warns Indy that nobody has ever left the cave alive. As proof, when Indy and Satipo enter the cave, they find the long-dead body of Indy's nemesis Forrestal.

Several terrifying booby traps later, Indy finds the Hovitos idol on an altar. Replacing the idol with a bag of sand to thwart any potential booby traps rigged on the altar, Indy takes off with the idol. The entire temple starts to collapse, and Indy and Satipo race from the destruction in a torrent of poison darts and arrows.

This whole opening segment is exciting and is a terrific way to get the audience revved up for an action-packed film. Now, let's take a look at the golden idol itself. Was there a South American tribe called the Hovitos? Did they possess golden idols and keep them in temples?

The idol that Indiana Jones snatches is sacred, according to the Hovitos tribe; so sacred, in fact, that they will kill him to get it back. In reality, the Hovitos tribe is probably based on the Chachapoyas people who lived in the Amazonian Andes, or Amazonas, region of northern Peru. The Amazonas is bordered by Ecuador to the north, and its capital is Chachapoyas, the name of the tribe. The Andes mountains are higher than any range outside of Asia, and it is somewhere in this mountainous region that the Chachapoyas dwelled, so high up that they were in the clouds. In fact, they were known as Warriors of the Clouds. The Chachapoyas spent their time hunting and foraging in the mountains' hot, moist evergreen forests.

Right before the Spanish came to Peru in the 1500s, the Incas had conquered the Chachapoyas. It is possible that the golden Hovitos idol is derived from the golden objects of the Incas, who were ruling the Chachapoyas. The capital of the Incan Empire was Cuzco. In the center of Cuzco was a great temple of the sun, Koricancha, which means "storage of gold" in the Quechua Incan language. Idols from all of the provinces that the Incans had conquered were stored in the Koricancha temple. Supposedly, there was so much gold in the temple that even its walls and floors were sheathed in gold. The courtyard of the sun temple was filled with golden statues, and the temple itself stored many sacred idols.

It is also possible that the Hovitos temple is based on the ancient Chachapoyas temple of Keulap, which is approximately two thousand feet long and is perched on the top of a ten-thousand-foot-tall mountain. The outer two walls of Keulap were made from stone blocks, each wall weighing ten thousand tons. To reach Keulap, people had to climb up a single-file hundred-foot-long passage with walls sloping inward. The Chachapoyas hoped that they would be protected from the Incans in this way, because they could attack the invaders from above while the Incans made their way up the narrow passage. To get to the Hovitos temple, Indiana Jones must climb up a similar narrow passage.

The Chachapoyas dominated northern Peru from 700 until 1480, when the Incan Quechuas conquered them. The Quechuas exist today in Peru, and they still speak their ancient Incan language. They wear woolen ponchos and caps made of bright colors and patterns, just as the Quechuas wear in *Raiders of the Lost Ark*.

As for the Chachapoyas, their ruins are well known among archaeologists. They built fortresses all along the mountains, and military strength was important to them. Because they fought so fiercely against the Incas for many years, the Chachapoya warriors gained a reputation for being brutal and aggressive.

Spanish soldiers arrived in Chachapoyas in 1547 and put the remaining natives into settlements, where poverty and disease took hold. Under Spanish rule over the next two hundred years, the population of Chachapoyas decreased by 90 percent.

By the time Indiana Jones reached Peru in 1936, the Chachapoyas were long gone, and archaeologists knew of the locations of very few Chachapoyan ruins. Their fortresses, idols, and mummies had sunk deep into the dangerous jungle beneath the dense clouds.

Deadly Tarantulas

When Indiana Jones and Satipo enter the Hovitos temple, three deadly tarantulas crawl up Indy's jacket. He easily swats them off using his trusty whip and then motions for Satipo to turn around. Satipo's back is covered with the spiders.

Most people are terrified of tarantulas. There's something about spiders in general that gives people the creeps. The bigger the spider, the scarier it is. Many of these fears are actually unfounded, because most spiders don't hurt humans. Of course, if you've ever been bitten by a spider, you might think differently. Spider bites are irritating and they itch.

The tarantula, a huge, hairy arachnid, is in the Theraphosidae family. It has feet, or tarsi, with two claws and tufts called scopulae. Approximately eight hundred species of tarantulas are known, and all hunt prey on the ground. Tarantulas eat mainly insects, although some large tarantulas also eat mice, birds, and lizards. Tarantulas look horrifying, but most types are not dangerous to humans, much less deadly.

Tarantulas are named, oddly enough, after Taranto, a town in Italy. The name *tarantula* originally described a species of European wolf spider. However, when European explorers in the New World encountered huge, hairy spiders unlike any they had seen before, they called them tarantulas and the name stuck.

Most tarantulas have a body size of approximately one to four inches long, and with their leg lengths taken into account the spiders range from three to thirteen inches long. The body length measurement is taken from the tip of a back leg to the tip of the front leg on the same side of the spider. The largest tarantulas weigh about three ounces and tend to hail from Brazil or Venezuela. For example, the Goliath birdeater tarantula, found in both countries, can have a full thirteen-inch body length and weigh three ounces. The pinkfoot Goliath may also have a thirteen-inch body length, and the Brazilian salmon birdeaters are equally as large.

When a tarantula eats, it secretes a digestive enzyme through its salivary glands and injects the fluid into its prey using fangs. The enzyme digests the prey from the inside out, turning all of the tissue into liquid. Then the tarantula sips the liquid out of what is left of the prey, which is typically an undigested shell.

Although the killer enzyme liquidates insects, mice, birds, and lizards, it does not kill humans. Instead, it produces pain and swelling. Some tarantulas may have chemicals on their abdomen hairs that cause human skin rashes and inflammation of the nasal passages and the eyes.

There are some dangerous spiders in the world that resemble tarantulas. These spiders are related to tarantulas, being in the same suborder, but they are not in the same family, Theraphosidae. It's possible that the deadly reputation of these other spiders, coupled with the image of a huge, hairy tarantula, provided the inspiration for killer-spider movies such as *Raiders* and *Arachnophobia*. The Brazilian wandering spider looks somewhat like a tarantula, being hairy and about five inches long, and its bite is highly poisonous to humans. The venomous Sydney funnel-web tarantulas, which aren't really tarantulas despite their name, are also extremely poisonous. Their bite is lethal and resulted in human deaths before an antidote was discovered in the 1980s.

Are there tarantulas in South America? Yes. In fact, some South American people actually roast tarantulas and eat them.

The *Avicularia* genus of the family Theraphosidae includes several species of South American tarantulas. When this type of tarantula is threatened, it first tries to jump or run away, but if it senses that it is under mortal attack, it sprays excrement at its predator. The excrement will hit the tarantula's opponent with great accuracy at a distance of two or three feet. So, rather than getting a death bite, Indiana Jones might have gotten sprayed with excrement. In the movies, however, fact is sacrificed to fiction because it is far more exciting to portray a death bite.

Gigantic Rolling Boulders

After stealing the golden Hovitos idol, Indiana Jones races from the temple. He is almost killed by a gigantic rolling boulder and, barely escaping, runs into his archrival, Rene Belloq.

At first thought, it seems improbable that boulders can be perfectly spherical. Gigantic, yes; spherical, no. But on second thought, the idea makes sense.

Examples of spherical boulders are found worldwide. Huge spherical boulders lie on the Koekohe Beach of the New Zealand Otago coast. Local legends claim that these Moeraki Boulders are the remains of kumara sweet potatoes and calabashes, which are gourds, and baskets that were used to catch eels. It is said that these remains washed ashore when a large sailing canoe called the Arai-te-uru was destroyed at sea.

Many of the boulders are 1.5 to 3 feet in diameter, and these are the smaller ones. A full two-thirds of the boulders are much bigger, ranging from 4.6 feet to 6.7 feet in diameter. Nearly all of the boulders are perfect spheres.

In Hokianga Harbour, North Island, New Zealand, you will find the Koutu Boulders. Some are as large as nine feet in diameter, and nearly all are entirely spherical. And the Katiki Boulders, approximately twelve miles south of the Moeraki Boulders, are also completely round.

In the United States, huge spherical boulders are located in North Dakota, where they can be as massive as ten feet in diameter. In Wyoming, Kansas, and Utah, enormous spherical boulders range up to eighteen feet in diameter

How are such immense round boulders created? In the case of the Moeraki Boulders, they are the cemented remains of Paleocene mudstone, formed by calcite precipitation. The boulders formed in the mud on the bottom of the Paleocene sea and took approximately

5 million years to grow. The calcite precipitation was caused by bacteria that reduced the sulfate of the saline within the mudstone. Large cracks appeared in the boulders, and, for the most part, these were filled in with brown and yellow calcites. The spherical shape was due to the mass diffusion of calcium, rather than a liquid flow over the boulder.

Giant boulders are found all over Peru, where Indiana Jones and the Hovitos have their adventures. In the jungles surrounding Cuzco and Machu Picchu, where the Chachapoyas and the Incas lived, huge spherical boulders lie in the rivers and in pools of swirling water. So it is very possible that a gigantic, spherical boulder could come crashing down after Indiana Jones tries to escape from the temple.

South American Indian Weapons: Poison Darts and Bows and Arrows

After forfeiting the idol to Belloq, Indiana escapes once again. But then Belloq orders the Hovitos to stop Indy, so they chase him and try to kill him using poison darts and bows and arrows.

South American tribes have long used both types of weapons in hunting. It is very conceivable that the Hovitos would use these against Indiana.

The Hovitos would make their darts the same way that actual South American tribes made theirs: from sharpened cane sticks, with either kapok tree fiber or cotton on the ends. The darts are pointed and cut like corkscrews on their tips, which are coated in poison. They are catapulted at prey via blowguns that are often made from cane. The shaft of the blowgun is long and fashioned from one piece of cane, and thinner pieces of cane are sometimes added to the large inner piece. Wood is used for the mouthpiece. Other Indians blow the darts out of reed shafts that are up to twelve feet long. Using this method, they can hit an animal with great accuracy from a distance of a hundred yards or so.

The poison that tips the darts is most likely gleaned from the glands of the poison dart frog, also known as the poison arrow frog, the poison frog, or the dart frog. This frog comes in several varieties, all belonging to the family Dendrobatidae. Nearly all poison dart frogs live in South and Central America, although a particular type, the *Dendrobates auratus*, has been seen on the Hawaiian Islands. The most poisonous of all the frogs is the golden poison dart frog, which lives primarily in Colombia.

The skin of the frogs contains poisonous alkaloids such as Batrachotoxin, which cause victims' muscles to contract so that the victims cannot move and their hearts fail. The poison is so potent that if a frog is in contact with a leaf or a piece of cloth, and an animal sits on the leaf or the cloth, the animal will die. Licking or swallowing a frog means certain death. The bright colors of the frogs help to keep predators away. A wild golden poison dart frog contains enough alkaloid poison in its skin to kill a hundred people. The poison is stronger than curare, which is a widely used jungle plant mixture that Indians east of the Andes use on their arrows and blowgun darts.

South American Indians catch the poison dart frogs in the jungle and keep them in hollow canes. When poison is needed, the frog is removed from the cane and a sharp stick is thrust down the frog's throat and straight through one of its legs. The tortured frog perspires from the pain and fear, particularly on its back, which becomes coated with an extremely toxic white poisonous froth. The Indians dip the points of darts into the froth. Darts prepared in this way maintain their ability to kill for an entire year.

Beneath the white froth is a poisonous yellow oil that the Indians preserve for later use. The oil remains toxic for six months.

Certain South American Indians have another way to get the poison out of the frog. They stroke the frog (the poison must enter the bloodstream to be deadly, so it is not absorbed through the skin) or warm it over a fire on a skewer to make it excrete large amounts of poison. After obtaining the poison, they boil it to make it extremely potent. Yet other tribes simply roll the tips of their darts across the backs of golden poison dart frogs.

The Indians carry the darts in quivers, typically made from bamboo, although the tops of the quivers are sometimes crafted from animal hide. Sometimes the quivers are even made from leaves.

The darts are so sharp and strong that they can penetrate tree trunks. Tipped in poison, the darts are potent enough to kill large animals.

The use of bows and arrows is also common. The bow is constructed from a flexible piece of wood and is strung with fiber. The Indians make their arrows from cane and feathers. To carve the arrowheads out of hard woods, the Indians use animal, fish, or bird bones or sharpened twigs. Different sizes of arrowheads are used for different prey—larger ones for larger mammals, smaller ones for birds and small mammals.

The arrows are carried in a quiver suspended by a strap hung around the neck, with the quiver hanging on the hunter's back between his shoulder blades. The strap is made of bamboo stalks so it does not cling to the neck.

Archaeology and Real-Life Archaeologists

After his adventure with the Hovitos idol, Indiana Jones returns to the United States, where he is known not only as a famous archaeologist but also as an expert on the occult who has a gift for finding rare antiquities. Marcus Brody, a friend of Indy's father, funds many of Indy's archaeological expeditions and calls on him for one more.

In real life, archaeologists study human culture by finding and analyzing the physical-material remains of past societies. Quite often, the only way we can begin to understand how humans lived in the past is to uncover these remains. Thousands of cultures have died out, leaving no written records. Even when we have written records, they are often misleading or incomplete. It is certainly true that our knowledge of the earliest human civilizations comes exclusively from archaeology, and this includes information about religions; the building of villages, towns, and cities; and the beginnings of agriculture.

Today's archaeologists do surveys before they initiate an expedition to a remote location, such as the deep jungles of Peru. They use surveys to find unknown sites, perhaps villages and homes. Indiana Jones, however, used much more direct methods: he journeyed to Peru using a tattered map and began his quest.

Modern excavation is also a bit different from Indiana's strategy for obtaining the Hovitos idol. While Indiana evaded ancient booby traps to uncover the idol, modern archaeologists determine the precise locations of the artifacts they want to uncover, and then they carefully dig to remove the objects without damaging them or any other relics in the area. Archaeologists keep accurate records of exactly how the artifacts are removed and how they are situated relative to other objects. In this way, an archaeologist can determine how humans used various artifacts together.

Precise record keeping is important because quite often the same site is used by one culture after another over the course of thousands of years. If an archaeologist were to remove artifacts without noting their precise locations, he or she would not have much idea about how the objects were used. Did an object come from an early society or a recent one? Was this object, found slightly lower in the ground, used in conjunction with an object unearthed closer to the surface? Typically, artifacts from earlier cultures are found beneath those from recent cultures. It is unlikely that the excavation of a golden idol would really be conducted by an archaeologist in the way that Indiana Jones does it.

Many experts claim (as Belloq does later in the film) that "archaeology is not an exact science." For example, when discussing the excavation of the Dead Sea Scrolls, Professor of Religion Robert Eisenman of California State University at Long Beach wrote:

> [P]aleography [the study of ancient writing and inscriptions] is being used now to authenticate all kinds of different objects. It was used on the Dead Sea Scrolls to make all sorts of extravagant claims that were often at odds with the internal evidence or what the texts themselves said. These are not exact sciences. Yet the public has been given the impression that they are, when in fact they are extremely questionable.[1]

Michael Kunz, another archaeologist, said, "We are a discipline, not an exact science. We shouldn't pretend we are. Everything is subject to interpretation."[2]

According to an article from Brown University, "Scientific Archaeology vs. The Discovery Channel," by Martha Joukowsky, early archaeologists were not particularly scientific, which resulted in their making serious mistakes. The first excavations were performed not by professionals, but rather by wealthy people with a lot of time on their hands. This upper-crust hobby became more of a science as it turned into a profession, but was not an exact science by any means. The reason is that entire layers of cultural artifacts were destroyed during excavations because the archaeologist was interested in only one particular layer. And even when early archaeologists were careful to preserve and record other layers of cultural materials, the records were often sketchy and the information was forever lost.[3]

In reviewing an archaeology book called *Landscapes of Change*, Paolo Squatriti of the University of Michigan discussed the opinions about archaeology and science of the various archaeologists who contributed to the book. He wrote:

> Christie notes that "landscape archaeology is not an exact science or discipline" . . . and its inexactitudes receive attention throughout. Guy Sanders probably goes furthest in this regard. . . . In his discussion of south Greek (mostly Corinthian) rural evidence, Sanders points out how the assumptions and cultural constructs of earlier scholars shaped a "catastrophist" account of settlement and economic activity that now seems untenable, especially in light of many artifacts' re-dating. Similar erroneous readings of the pottery supply are deconstructed in Leone and Mattingly's chapter about Maghribi contexts.[4]

These are only a couple of examples of how archaeology has failed to be an exact science throughout the years.

Yet in modern times, archaeology has matured into a more exacting occupation. Now professionals require precise measurements to record data. Indiana Jones, however, is more like an archaeologist of the past, who uncovered artifacts of interest without taking accurate measurements or noting additional data for future use.

A Stanford University newsletter reported that "[A]rchaeologists, in this view, do not offer an exact translation of the past for their contemporaries; rather, they are mediators."[5] Along these lines, entire books have been written about archaeology and how it borders on folklore. College courses focus on the subject, and various museums around the world are devoted to archaeology and folklore. After all, ancient artifacts are usually saturated in folklore and religious overtones because the local people are attempting to understand the remains of their own past.

Folk stories connected to artifacts have recurring themes that often relate to religious entities or beings that inhabit the artifacts. For example, many cultures have believed that ghosts, ancient rulers, devils, and religious entities are intimately connected to their ancient artifacts.

Were there really archaeologists like Indiana Jones, the pure adventurer? It's extremely likely that adventurers have gone deep into the Peruvian jungles seeking their fortunes.

Notable individuals who resembled Indiana Jones include Senator Hiram Bingham III, who excavated Machu Picchu in 1911 and wrote a book about his discoveries called the *Lost City of the Incas*.[6] Machu Picchu was a major settlement of the Andean Indians and was thought to be a stronghold retreat for the Incan rulers. In the Quechua language, the name of the settlement literally means "old mountain." While others contributed to the exploration of the Lost City, history identifies Hiram Bingham as its discoverer.

Alfred M. Bingham wrote a biography of his father in which he said that Hiram Bingham had been brought up in a family of missionaries.[7] Rather than continue the tradition of religion and poverty, he instead decided to seek a career in the ivory towers of academia. He married a granddaughter of the founder of the Tiffany company and hence came into a vast amount of wealth. He used his wife's riches to help fund five expeditions, one of which was the exploration that led to the discovery of Machu Picchu.

Bingham was born in Hawaii, where his family served as Protestant missionaries. He went to school in Hawaii from 1882 to 1892, then graduated from Phillips Academy in Andover, Massachusetts, in 1894. His college degrees were from Yale University, the University of California at Berkeley, and Harvard University. After completing his education, Bingham taught history and politics at Yale, spent some time at Princeton, and ended up in 1907 as a lecturer on South American history at Yale, in Connecticut. He later became a professor there. Note that Indiana Jones was also an authority on South American history and served as a lecturer-professor at Marshall College in Connecticut.

In 1911, Bingham traveled to the Andes with the Yale Peruvian Expedition. A local policeman, Agustín Lizárraga, helped Bingham to find Machu Picchu, which had been discovered previously but then long forgotten. According to *National Geographic*,

> [i]n 1911 Yale University Professor Hiram Bingham, searching for the lost Inca capital of Vilcabamba, paid a Peruvian guide to lead him to a nearby ruin. The guide took him 2,000 feet (610 meters) up a precipitous slope—and straight into the "lost" city of Machu Picchu. . . . Arguably the greatest archaeological site in the Americas, Machu Picchu remains a mystery. Some scholars believe it to be the birthplace of the Inca Empire. Others see a ceremonial center or military citadel.[8]

Nobody seems to know how the civilization that built Machu Picchu ended. The Spanish have no records of its existence.

Although Indiana Jones had a scrap of a map to guide him, Bingham had some seventeenth-century writings. And where Indy had two local guides, Bingham had one local policeman as his guide. He climbed a steep path, just as Indiana Jones did, and the similarities continue, as Bingham discovered an ancient temple over a cave.

In addition to his wife's support, Bingham received funds from Yale University, where he worked, and the National Geographic Society, both of which supported expeditions to Peru in 1912 and 1915. Hiram Bingham received the first National Geographic Society archaeological grant.

Indiana Jones was a famous adventurer, but he was also well known for his ability to locate and secure rare artifacts that were worth a fortune. In this way, Indy was also very similar to Hiram Bingham. During Bingham's adventures in Machu Picchu, he excavated thousands of valuable artifacts, which were later housed in Yale's Peabody Museum. In fact, in 2005, the Peruvian government threatened to sue Yale University unless the artifacts were acknowledged as belonging to the Peruvian people and returned. The issue remains unresolved.

Another archaeologist on whom the character Indiana Jones may have been based was British colonel Percy Harrison Fawcett. This adventurer disappeared with his son in the Amazon jungle in 1925 while searching for a lost city he called "Z."

Fawcett was born in 1867. His father was a fellow of the Royal Geographic Society, which gave Fawcett an early interest in exploration and adventure. Percy Fawcett served in the Royal Artillery and the British secret service in North Africa, where he learned how to be a surveyor, a key skill for early adventurers seeking lost cities. Of interest to science-fiction and mystery fans, Fawcett's friends included Sir Arthur Conan Doyle, the author of the Sherlock Holmes tales, and H. Rider Haggard, who wrote many adventure novels that were extremely popular in their time.

In 1906, on behalf of the Royal Geographic Society, Fawcett headed to South America to map the jungle along the border of Brazil and Bolivia. Between 1906 and 1924, he made seven exploratory trips in South America.

Then in May 1925, Fawcett sent a telegraph to his wife that he was going into unexplored territory and might not be seen again. With one companion, Raleigh Rimmell, Fawcett and his son set off to find what Fawcett believed to be a lost Amazon city. It is thought that they were somewhere near a southeastern tributary of the Amazon River when they disappeared into territory occupied by several South American Indian tribes. Some people assumed that the Indians had captured and killed Percy Fawcett. Others suggested that wild animals killed him. Yet others went so far as to claim he became the chief of a tribe of cannibals deep in the jungle. Thirteen expeditions were launched to find the adventurer and his party, but the results were always the same: Percy Fawcett was lost forever.

It's conceivable that the Hovitos tribe that attempted to kill Indiana Jones was loosely based on the tragedy surrounding Percy Fawcett and his quest to find Z, the Amazonian lost city.

Vendyl "Texas" Jones, born in 1930, is yet another archaeologist and adventurer whom Indiana Jones resembles. In fact, Vendyl's first name could be shortened to Vendy, which is very similar to Indy, and, of course, his last name is Jones. Plus, his nickname, "Texas," is the name of an American state—as is "Indiana."

Jones is a religious man—unlike Indiana—and received degrees in divinity and theology from the Bible Baptist Seminary. His studies continued at the Bowen Biblical Museum under the tutelage of Dr. and Mrs. William Bowen, as well as with a biblical archaeologist, W. F. Albright.

While serving as a Baptist pastor, Jones realized that the anti-Jewish comments in the gospels were probably *not* included in more ancient manuscripts. He called a rabbi and began what became a lifelong pursuit of biblical truth. In particular, Jones was interested in finding the original sources of biblical material and other ancient religious materials. Jones continued an earnest study of Judaism and later established the Judaic-Christian Research Foundation. The parallel between Vendyl Jones's quest and the search for the Ark of the Covenant (discussed later in this book) by Indiana Jones is a reasonable one.

In 1964, explorers from the Jordan Department of Antiquities found the Copper Scroll, one of the Dead Sea Scrolls, in Cave #3 at Khirbet Qumran, Israel, on the northwest shore of the Dead Sea. The Dead Sea Scrolls are a collection of approximately 850 ancient articles. Authorities place the authorship of the Dead Sea Scrolls between the middle of the second century BC and approximately AD 100. Some of the Dead Sea Scrolls were recorded on papyrus, while many were written on brown animal hide. The Copper Scroll, as its name implies, was inscribed on thin sheets of copper mixed with approximately 1 percent tin.

Of interest to Indiana Jones aficionados is that the Copper Scroll identified the hiding places of sixty-four sacred articles. Some people say that both the Tabernacle and the Ark of the Covenant were on the list, as well as the Holy Incense and the Holy

Anointing Oil. What we do know is that the Copper Scroll lists the hiding places of more than a hundred tons of gold and silver items.

After two thousand years in a cave, the scroll was badly oxidized and clearly would have crumbled if anyone had attempted to unroll it. Four years after its discovery, the Copper Scroll was sent to Manchester College of Technology in England, where it was opened in such a way as to preserve its contents. Indiana Jones would have found the Copper Scroll to be of great interest. And so, as it happens, did Vendyl "Texas" Jones.

In 1967, Jones moved to Israel to continue his intensive study of Judaism at Hebrew University, and there he became fascinated by Israeli archaeology. After the Six-Day War, he joined an excavation team en route to Qumran, where the Copper Scroll had been found. He later went on many archaeological expeditions to Qumran, with more than three hundred volunteers helping him. He received no funding from governments or foundations, and, indeed, the Israeli government did not even provide him with digging permits.

The parallel between Indiana Jones and Vendyl "Texas" Jones is tenuous, at best—except for their names. Although both Joneses sought Judaic relics, one (Texas) was a religious devotee, and the other (Indy) was simply an adventurer with a knack for unearthing extremely rare artifacts.

In 1988, Texas Jones's foundation reported that one of its excavation teams had found some Holy Anointing Oil from the original Holy Temple. Then in 1992, the foundation claimed it had found Holy Incense, and the Weizmann Institute of Science is reported to have analyzed the findings as containing eleven ingredients of Holy Incense. The biblical ingredients were spelled out by God to Moses in the book of Exodus and included stacte drops, onycha, galbanum, and frankincense.

Many other adventurers had lives that make the career of Indiana Jones plausible. For example, Roy Chapman Andrews led expeditions into the Gobi Desert and Mongolia in the early twentieth century and found fossil dinosaur eggs for the first time. His adventures included near-death experiences and harrowing escapes from armed Chinese bandits, sharks, and pythons.

Another example is Robert Braidwood of the University of Chicago, who led an expedition to the Amuq Plain in Hatay, Turkey, and performed one of the first scientific archaeological surveys. And archaeologist Sylvanus Griswold Morley performed extensive excavations of the Mayan site of Chichen Itza, a huge pre-Columbian site in the northern center of the Yucatan Peninsula.

There are, no doubt, countless other examples of adventurers seeking golden idols and religious artifacts in the style of Indiana Jones.

Bullwhips

While escaping with the golden Hovitos idol, Indiana and Satipo are separated by a deep hole in the ground, with Satipo on one side holding Indy's whip. Indy tells Satipo to throw him the whip so he can swing across the chasm. Satipo responds, "Throw me the idol, I give you the whip." So Indiana throws the idol to Satipo and waits for Satipo to toss him the whip. Satipo drops the whip to the ground, replies, "Adiós, señor," and absconds with the idol.

In *Raiders of the Lost Ark*, Indiana Jones is rarely without his bullwhip. The bullwhip, made from leather (hence, the term *bull*whip), is one of mankind's earliest weapons. In the Indiana Jones movies, Harrison Ford used bullwhips that were up to ten feet long, the length depending on the particular stunt. They were owned by stunt coordinator Glenn Randall and were made by David Morgan, a famous maker of whips.

In reality, Harrison Ford did not use a bullwhip to swing across the chasm in *Raiders of the Lost Ark*. Rather, it was a steel cable wrapped in braided leather. A bullwhip will snap from the weight of a man's body being swung across a chasm.

Movie adventurers often use bullwhips because they make the heroes looking dashing and courageous. A familiar example is Zorro, created in 1919 by Johnston McCulley. In the pulp magazine *All-Story Weekly*, Zorro made his first appearance in "The Curse of Capistrano." He is famous for his attire and his weapons: black clothing, a black Spanish cape, a black flat-brimmed hat, a black cowl mask covering the top of his head, a rapier with which he cuts his distinctive Z mark, and, of course, his bullwhip. It was this dashing, courageous feel that the makers of *Raiders of the Lost Ark* wanted, so they supplied Indiana Jones with his bullwhip. Later, in the other Indiana Jones films, David Morgan was hired to custom make dark-brown bullwhips for Indiana Jones.

Other than their excellent use as movie adventurers' weapons and tools of trade, bullwhips have more practical uses. Traditionally, animal handlers use them to control livestock. The design of the bullwhip enables it to make a little snapping sound at the end of the throw, when part of the whip slightly exceeds the speed of sound.

Of interest to Indy fans is that the origin of the bullwhip may have been in South America. Roman mosaics from AD 2 show tiny one-piece whips, however, so the South American origin remains debatable. The Spanish vaqueros, the original cowboys, introduced their bullwhips into Mexico, and from there, people brought the whips further north into the United States.

Fedoras

Just as Indy is about to be overtaken by a pursuing band of Indians, he jumps into a river and swims to a waiting seaplane. His hat never falls off.

While we're thinking about Indiana's bullwhip and its origins, let's ponder some of his other accoutrements—specifically, in this section, his trademark fedora. According to film lore, Steven Spielberg and George Lucas both wanted Indiana Jones to have a distinctive style about him, something that moviegoers would never forget. In particular, they were interested in giving Indy a really cool hat.

The hat became part of Indiana Jones's iconic look, and Spielberg and Lucas got what they wanted. The hat chosen for Indiana Jones was a wide-brimmed, tall-crowned fedora. This type of hat has become associated with a rugged, cowboylike persona and was worn by real adventurers, as well as by adventure-movie heroes, long before Jones went to Peru.

In the Tarzan films of the 1930s, the white explorers wore safari hats, which looked similar to medium-brimmed fedoras. In fact, depending on the scene, Indiana Jones's hat sometimes looks more like a safari hat than a fedora.

Although it often resembles a safari hat, Indiana's fedora leans on the side of dressy fedoras rather than outdoorsman-type gear. Yet this fancy version of a fedora also has roots in early adventure films, such as *Secret of the Incas* from Paramount Pictures in 1954. This low-budget adventure release starred a young Charlton Heston as Harry Steele, whose goal was to find ancient Incan treasure in Machu Picchu. Very similar to Indiana Jones, Harry Steele uses a stone map excavated from Machu Picchu to try to hunt down the location of the solid-gold "Sunburst" treasure. Not only does Harry Steele wear an Indy-like brown fedora, he also wears a leather jacket, something else that Indy is famous for.

After *Raiders of the Lost Ark*, a different hat was worn by Indiana Jones. The *Raiders* hat had a tall crown, and the fedora in *Temple of Doom* had a shorter crown.

Leather Jackets

Like Indiana Jones's famous fedora, his leather jacket has long been associated with him, but it traces back to early adventurers. Indiana's jacket was a facsimile of the leather jacket styles of the 1930s. These were early versions of the A-2 jackets worn by American pilots in World War II.

As we noted previously about the fedora, Harry Steele wore a leather jacket similar to Indy's jacket in the film *Secret of the Incas* from Paramount Pictures in 1954. Another example of the fedora-and-leather-jacket look is seen in the movie *China*, also from Paramount and released in 1946.

Real adventurers have worn leather jackets since the cowboy era, but their use became widespread when aviators started to wear them after World War I.

Deborah Nadoolman designed Indiana Jones's leather jacket, which was made by Berman's and Nathan's in London. Peter Botwright, the owner of Wested Leather in London, manufactured the *Last Crusade* Indy jackets, which were constructed in a slightly different manner from the previous ones. For example, in *Indiana*

Jones and the Last Crusade, the jacket's storm flap has a snap button to keep it closed.

The Nazis and the Occult

After Indiana's adventure with the Hovitos idol, he returns to his professorship at Marshall College in Connecticut. Two U.S. Army Intelligence officers show up at the college and hire him to find the long-lost Ark of the Covenant, which Army Intelligence has learned the Nazis are hunting. The Nazis are seeking occult power and want to find Abner Ravenwood, who served as Indiana Jones's mentor before Indy got his PhD. Ravenwood is the world's expert on Tanis, an ancient Egyptian city, where the Nazis are digging for the Ark. The army believes that the Nazis want Ravenwood because he owns the headpiece to the Staff of Ra, which is key to calculating the exact location of the Ark.

We will talk a bit about all of these topics—the Ark of the Covenant, Tanis, and the Staff of Ra—later in this book, but for now let's focus on the Nazis and their obsession with the occult.

Many high-ranking Nazis, such as Rudolf Hess, Heinrich Himmler, and Richard Walther Darre, were interested in the occult, and Hitler believed that he himself was a godlike creature. Hitler claimed that during World War I, he heard a voice warning him to leave a crowded dugout just before a shell fell on the spot, destroying the bunker and killing all of its inhabitants. He believed that his experience with the voice and the shell was an indication that he had a special role to fulfill in the world. In fact, he became obsessed with the notion that a heavenly force was protecting him.

Also during World War I, a highly decorated British soldier named Private Henry Tandey had a clear shot at Hitler but passed on the opportunity to kill him. Tandey had a moment of pity and empathy for the young man on the other side of the war, and because of Tandey's kindness, the world suffered greatly. Oddly enough, during World War II, Tandey was the most highly decorated soldier in the British Army. He received a Victoria Cross for his bravery during the Marcoing battle, which is where he lowered his rifle and let the future Führer go. Hitler witnessed Tandey lower his rifle and decided that the gods of war had come down to Earth and saved him. In memory of this holy moment, much later in 1937, Hitler requested a painting of Tandey, then hung it on a wall at Berchtesgaden.

There are people who believed that Hitler was possessed by demons, but there is no proof of this claim. Hermann Rauschning, a top Nazi aide and a fairly unreliable source of information, wrote a book in which he claimed that Hitler was possessed. Pope Pius XII performed exorcisms on Hitler—from a distance—three times. And Pope Benedict XVI thought it was possible that Hitler was possessed by demons. Hitler did not believe that demons possessed him; he thought that the gods protected him and had a spiritual role for him to play in the world.

But the Nazis' belief in mysticism and divine intervention on their behalf goes much deeper than Hitler's obsession with himself as having godlike power. Most of the German public believed his myth, too. In fact, to Nazis, Hitler was another Jesus, idealized as a savior from God.

Heinrich Himmler, another Nazi top aide, was fascinated by forms of Germanic neopaganism, as well as by the obvious Aryan racism. Neopaganism was a modern form of paganism and included animal sacrifices. The deities of Germanic neopaganism included Anglo-Saxon and Norse gods. Adherents of the religion also worshipped ancestors, viewing the gods as their progenitors. In addition, Germanic neopaganism included rituals dedicated to creatures such as dwarves and elves. Less popular forms of Germanic neopaganism included Seid and Spae, which involve sorcery and witchcraft, shamanism, and the telling of prophecies.

Himmler, thinking that he was the reincarnation of Heinrich the Fowler, established a philosophy called Esoteric Hitlerism. Heinrich the Fowler happened to be the founder and the first king of the medieval German state. Until Heinrich the Fowler's time, Germany was known as East Francia. Heinrich the Fowler was the Duke of Saxony from 912, and then king of the Germans from 919 until he died in 936. Thus Himmler, thinking himself a reincarnation of the first king of Germany, organized SS rituals and had his Wewelsburg castle private quarters decorated in honor of the king.

It must be obvious by now that certain high-ranking Nazis were indeed interested in branches of the occult. But things are even stranger than you think. In 1935, Himmler founded something called the Ahnenerbe Society, which not only focused on providing evidence that the Nazis were superior to all other people but also focused on the occult. Mystical organizations, populated by SS members, proliferated during the Nazi era. In fact, the SS had an occult unit. The Nazis would stop at nothing to prove how superior they were. Among their many practices, some of which we will describe briefly, they organized an expedition to Tibet to find the origins of the Aryan race.

One expedition organized by Himmler went to Finland to research pagan sorcerers and witches. Pagan chants were recorded, and illustrations were drawn of pagan rituals.

Central to the beginning of *Raiders of the Lost Ark* was the notion that the Nazis might send an expedition to the Andes: in reality, they did just that. In addition, they went hunting for the Holy Grail, said to be present during Christ's Last Supper. So the Nazis, in reality, were not only interested in the occult, they were actively searching for rare religious artifacts. The suggestion in *Raiders* that the Nazis might be looking for the Ark of the Covenant is not at all far-fetched. They actually did. They just never found it.

As noted, the driving force behind the occult in Nazi Germany was the Ahnenerbe Society. The society operated in concentration camps, using prisoners as subjects in a wide array of inhumane torture experiments. Prisoners were placed in tanks of freezing water with electrodes on them. They were given a substance called Polygal, which supposedly would coagulate blood to heal gunshot wounds, then they were shot so that the Nazis could determine whether the Polygal worked. These are just a couple of the "minor" experiments that were performed; we will leave the more gruesome and horrific details for writers of other, more serious volumes of historical work.

Oddly enough, although Himmler and other high-ranking Nazis researched and believed in neopaganism and occult worship of various kinds, they instituted harsh punishments for people who were occultists. Many of these people ended up in concentration camps.

Another odd twist to the sick Nazi story is that Himmler actually had a personal occultist, Karl Maria Wiligut.

The SS employed many occultists to help them fight the war. Ludwig Straniak, Dr. Wilhelm Gutberlet, and Wilhelm Wulff all gave advice to the Nazis Walter Schellenberg and Himmler. The astrologer Wilhelm Wulff was told to find Mussolini, who was hidden fifty miles south of Rome, and Wulff used astrology and pendulum dowsing to get the answer. Using his psychic gifts, Wulff actually located Mussolini on the island of Ponti, where he was being held by Allied troops. The architect Ludwig Straniak was told to find a battleship, which was then at sea on a secret Nazi mission. Straniak dangled his pendulum over a map and located the battleship near the coast of Norway.

The Ark of the Covenant

Marcus Brody explains to Indiana that the Nazis want the Ark of the Covenant to make their forces invincible; legend has it that the Ark's supernatural powers can destroy entire armies.

The goal of Indiana Jones in *Raiders of the Lost Ark* is to find the long-lost Ark of the Covenant. If the Ark of the Covenant can destroy armies and make the Nazis invincible, it makes sense that the *Raiders* villains are willing to do anything to obtain the ancient religious object. In this section, we explore the powers of the Ark of the Covenant and see whether there's any basis to the Nazis' belief that the Ark can grant supernatural powers. But first, what exactly is the Ark of the Covenant? Did it really carry the pieces of the Ten Commandments that were broken by Moses? Let's find out what the Ark is, where it came from, and where it might be today.

Many religions venerate physical images of gods, whether they be in the form of statues (possibly golden idols), paintings, photographs, or other manifestations of spirituality. Judaism differs in this respect, absolutely rejecting any worship of physical manifestations of God.

The roots of Judaism go back to Abraham, who lived between 1813 BC and 1638 BC. (Yes, he lived to a ripe old age.) Born under the name Abram in the city of Ur in Babylonia, Abraham was the son of an idol merchant named Terach. As Abraham grew up, he questioned Terach about idol worship. Abraham came to the conclusion that there was a single Creator of the entire universe. When Terach left his idol shop in Abraham's hands one day, the boy smashed all but one of his father's wares with a hammer. He placed the hammer in the hand of the one remaining idol. Then he lied and told his father that the idols had all gotten into a huge fight, and the biggest one smashed all of the smaller ones with the hammer. Terach replied that the idols had no life or powers, and it was impossible that the large idol had smashed all the smaller ones. Abraham asked his father the obvious question: why believe the idols can do anything for mankind if they have no powers or life of their own?

Regardless, one day, the Creator whom Abraham worshipped called out to him with an offer. Abraham was to leave his family and home, and if he did this for God, then Abraham would become the leader of a great nation. Abraham accepted God's offer, and hence was born the *b'rit*, or covenant, between God and the Jewish people. In Genesis 12 of the Bible, God tells Abraham, "Get thee out of thy country, and from thy kindred, and from thy father's house, unto the land that I will show thee. And I will make of thee a great nation, and I will bless thee, and make thy name great; and be thou a blessing." Abraham was allowed to bring some family members with him, as Genesis tells us: "So Abraham went, as God had spoken unto him; and Lot went with him. And Abraham took Sarai his wife, and Lot his brother's son, and all their substance that they had gathered, and the souls that they had gotten in Haran; and they went forth to go into the land of Canaan."

The b'rit is a fundamental aspect of Judaism, an acknowledgment that the people have obligations to God, and that God has certain obligations to the people. Eventually, God gave the descendents of Abraham the Torah, but first Abraham had to prove himself by performing ten tests of faith. The Jews became known as the People of the Covenant, the Chosen People, because they had made a covenant with God and thus had been chosen by God to be his representatives on Earth.

The Torah is also known as the Written Law. The first part consists of the Hebrew Bible, known commonly by non-Jews as the Old Testament, which includes the section called Nevi'im, or Prophets; and the second part is called Ketuvim, or Writings. Other names for the Torah are the Five Books of Moses, the Pentateuch, or, if it is not in written in scroll form, the Chumash. And for Orthodox Jews, the term *Torah* might even include aspects of later writings, known as the Oral Law, such as the Mishnah, the Midrash, and the Talmud.

The Torah condemns the worship of idols. In Exodus 32, God is furious with the Israelites because they construct a golden calf, which they hope will be an intermediary between the people and God. But in the time of very early Judaism, one man-made construction was considered holy. That was the Ark of the Covenant. It was created while the Israelites were wandering with Moses through the desert, and they used it for worship until the First Temple was constructed. To the early Jewish people, the Ark of the Covenant was the one physical manifestation of God on Earth. It served as a constant reminder of their bargain, their covenant, with God.

According to Exodus 25 and 37, while the Israelites were camping at Sinai, God commanded Moses to build the Ark. The dimensions are spelled out in detail in the Hebrew Bible. With one cubit equaling approximately eighteen inches, God told Moses to make a box that was 2½ cubits long, 1½ cubits high, and 1½ cubits wide. The people made the Ark from acacia wood and lined it, on both the inside and the outside, with pure gold. In addition, gold rings were attached to the bottom of the box, and through these rings, the people slid two poles made from acacia and covered in gold. Using the poles, the family of the tribe of Levi carried the Ark on their shoulders. A gold covering, 2½ by 1½ cubits, was wrapped around the box, and two gold cherubs were attached to this covering. The cherubs faced each other and had wings wrapped around their bodies, with the wings of the two cherubs touching.

So this is what the Ark of the Covenant looked like and what it was made of: wood and a lot of pure gold. But what was in it?

That question has been the source of many debates throughout history. Many biblical scholars believe that the Ark contained the first tablets of the Ten Commandments that Moses brought down from Mount Sinai. These first tablets were broken by Moses, and he had to return to the top of Mount Sinai and ask God for a second set of tablets. Many scholars believe that the Ark also contained this second set of the Ten Commandments.

The Ten Commandments, of course, are central to many religions in today's world. They are essential to mankind's never-ending attempts to live in peace and harmony and to create societies that tame the more violent and unethical aspects of human nature. The Ten Commandments are the keys to humane, ethical, and moral conduct, and they were housed in the Ark of the Covenant.

The first two Commandments have special meaning for people of the Jewish faith: first, I am the Lord thy God, Who brought thee out of the land of Egypt and out of the house of bondage; and second, Thou shalt have no other gods before Me. Literally, there is one and only one God. The other Commandments are: third, Thou shalt not take the name of the Lord thy God in vain (in other words, do not use God's name when you are cursing or just in idle conversation); fourth, Remember the Sabbath day to keep it holy (this is Saturday in the Jewish religion, and, traditionally, Christians hold Sunday as their day of rest); fifth, Honor thy father and mother; sixth, Thou shalt not murder; seventh, Thou shalt not commit adultery; eighth, Thou shalt not steal; ninth, Thou shalt not bear false witness against thy neighbor; and tenth, Thou shalt not covet anything that belongs to thy neighbor.

Another holy object that was built around the time of the Ark was the Tabernacle. It was a portable temple that the Israelites carried with them in the desert. The Ark of the Covenant was stored in the Tabernacle, which also held other items used in religious worship. In Exodus 31:18, it is explained that God commanded the Israelites to build the Tabernacle because after the golden calf incident he realized that they needed some sort of physical representation of him, some physical mode of worship. But in Exodus 25:1, God commands them to build the Tabernacle *before* they commit the sin of the golden calf, so, possibly, God wanted them to use the Tabernacle to constantly remind themselves in some physical way of God's presence. Clearly, what God wanted is unknown, and these are the conjectures of biblical scholars.

While traveling in the desert, the Israelites carried the Ark two thousand cubits ahead of the people, as detailed in the Hebrew Bible. Surrounding this fairly straightforward comment are many mystical stories, passed down through generations. For example, in one such story, the bottom of the Ark of the Covenant shot two jets of fire into the path preceding the people, and the fire burned up scorpions and snakes. Another story claims that the Ark carried the people across the desert, rather than vice versa.

Given powers of these dimensions—shooting fire and carrying people across the desert—it is possible that the Ark could also destroy armies. In actuality, scholars have long debated whether the Ark helped the Israelites in fighting their enemies, or whether it had only a symbolic presence. If the Ark is at all magical, then who is to say where its powers begin and end? But first, of course, you have to believe the mystical stories.

If the Nazis in *Raiders of the Lost Ark* were anything like the Nazis of real life, then they might indeed, given their bent toward the occult, believe that the Ark of the Covenant could wipe out entire armies. One wonders, though, why the Ark would do this for Nazis. After all, the Ark is one with the Jews, not with those who seek to slay them.

One thing is clear: the Ark of the Covenant represented the physical manifestation of God. In Numbers 7:89, it is said that God spoke to Moses from between the two cherubs of the Ark, and he was accompanied by glowing clouds (per Exodus). While the Israelites wandered with Moses through the desert, clouds accompanied the Ark's movement, and these clouds connoted the presence of God. It is also written that clouds accompanied the Ark itself as the Israelites made their way through the desert, and at night, the clouds became a pillar of fire.

When the Ark was placed in the Tabernacle, and later when it was in the temple, people had access to it only once each year, on Yom Kippur, the Jewish Day of Atonement. Then, only the high priest could approach on that one day. According to Leviticus, the high priest addressed the Ark of the Covenant, asking forgiveness of sins for himself and all of the Jewish people. He did so in a thick cloud of incense.

The idea in *Raiders of the Lost Ark* that the Ark could be dangerous and could help its owners to destroy entire armies was probably based on all of these mystical stories surrounding it. In Leviticus in the Old Testament, it is mentioned that the Ark was dangerous: when Aaron's sons, Nadav and Avihu, wanted to offer a sacrifice in the Tabernacle, they brought a foreign flame into the Ark's presence; in response, the Ark consumed them with fire and burned them to death. The Bible indicates that this fire was created and sent by God.

Furthermore, when the Philistines captured the Ark, many people who came near the Ark or who simply looked at the Ark were instantly killed by its power. In Numbers 4:20, it was written that should a priest of the Tabernacle or the Temple look at the Ark when he wasn't supposed to look at it, the Ark would immediately kill him.

The Ark had many other powers that often saved the Jewish people, both in times of war and when they were escaping from their enemies. When the Israelites had to cross the Jordan River into Canaan, the Ark parted the waters and led the people through the resulting passageway. When the Israelites wanted to enter Jericho, they did so by marching around the walls, carrying the Ark and blowing horns.

After the episode at Jericho, the people set up the Tabernacle and the Ark in Shiloh, where they stayed until the Philistines went to war with them. The Israelites took the Ark from Shiloh, escaping a fierce battle that the Philistines won, but the Philistines ended up capturing the Ark. Upon hearing that the Philistines had taken the Ark of the Covenant, the high priest Eli immediately died.

In the south of Canaan in their capital of Ashdod, the Philistines put the Ark in the temple of their god Dagon. The mystical and supernatural powers of the Ark took hold once again, and the day after stealing the Ark, the Philistines found the idol of Dagon on its stomach, fallen from its pedestal. The following day, Dagon had no head. A short time later, Ashdod was hit by a terrible plague.

The Philistines tried to move the Ark from city to city, in hopes of protecting themselves from its powers. But wherever they took the Ark, the plague followed them, killing the occupants of their cities.

Finally, after seven months of hell, the Philistines returned the Ark to the Israelites. According to stories handed down through the generations, when the Israelites were once again in possession of their Ark, the ox that was pulling it home started to sing.

The Ark traveled quite a bit after this tragedy with the Philistines. It went from Beit Shemesh to Kiryat Yearim, and after twenty years, it went to Jerusalem. After David's son Solomon built the First Temple, the Ark was moved into the temple.

The temple was a beautiful construction. In one room was the Ark with its two tablets of the Ten Commandments. In 586 BC, approximately four hundred years after the temple had been built by Solomon, the Babylonians, led by Nebuchadnezzar, destroyed it, and the tablets of the Ten Commandments disappeared.

A second temple was constructed in the same location seventy years later, but there was no Ark to put in it. And then in AD 70, the Romans destroyed the second temple.

Most historians and biblical scholars do not believe that the Babylonians stole the Ark of the Covenant when they destroyed the first temple. The Babylonians kept detailed lists of everything they stole from the temple, and those lists did not include the Ark. There is always the possibility, however, that thieves took the Ark and melted it down for its gold.

Some sources claim that King Josiah of the Jewish kingdom of Judah hid the Ark before the Babylonians could find it and steal it. There are those who say that Josiah hid the Ark in a cave by the Dead Sea. Others say that he put it in a hole beneath the temple. If the Ark is beneath the temple, it may never be excavated, because both Israeli and Muslim officials would probably forbid any such excavation. Some people claim that the Copper Scroll, which we mentioned earlier in this book, describes where to find both the Tabernacle and the Ark of the Covenant.

According to the official Indiana Jones Web site, the film postulates that the Egyptian pharaoh Shishak (Sheshonq I) stole the Ark shortly after Solomon's reign ended. Shishak put the Ark in a vault called the Well of Souls, back in his capital city of Tanis in the Nile Delta. God destroyed Tanis with a yearlong sandstorm to punish Shishak for stealing the holy Ark.[9] How likely is it that the Ark of the Covenant is hidden in Egypt, as the Nazis think in *Raiders of the Lost Ark*? There is speculation that a group of Jews took the Ark to Tanis, exactly where the digs take place in *Raiders of the Lost Ark*, when they migrated to Egypt in the seventh century BC.[10] But these, like most theories, remain unproven.

Today, Jews keep what they call a Holy Ark in each of their synagogues, and inside the Holy Ark are the Torah scrolls. The Ark and the covers of the scrolls often have copies of the tablets of the Ten Commandments on them. The real Ark of the Covenant, however, has been lost for more than two thousand years.

The Staff of Ra

After accepting the job of finding the Ark of the Covenant, Indiana travels to Nepal in hopes of finding the headpiece to the Staff of Ra. This headpiece has a carving of the sun at the top.

In Nepal, Indy arrives at a bar called the Raven, which happens to be run by his ex-girlfriend Marion, the daughter of Abner Ravenwood. Abner Ravenwood is dead, but Marion has the headpiece. Of course, the Nazis are also in pursuit of Marion's treasure, now worn around her neck as a medallion. Indy has a brutal gunfight with the Nazi major Toht, ending with the Raven burning to the ground. Together, Indy and Marion race off to Cairo, Egypt, with the headpiece.

First, there was indeed an Egyptian god named Ra, sometimes called Re, and he was the patron of the sun, of light, of power, and of the pharaohs. Ra was associated mainly with the midday sun, which is the hottest sun of the day, and he looked like an Egyptian pharaoh with the sun disk on his head. For centuries, Ra was the main sun god throughout Egypt. Worship of him was based somewhat in Heliopolis, the City of the Sun. The sun was thought to be Ra's body or his eye. In addition, Ra was king of the gods and the creator of all, and mankind was created from his sweat and tears. Later, he was known as Amun-Re, meaning literally "the sun."

Along with Osiris, Ra did not live on Earth. Instead, the Egyptian god Horus ruled over the Earth. Ra was powerful but aging, so he remained in the skies to rule over people since he couldn't deal with them very well anymore. To represent his aging, Ra was sometimes shown in pictures as an infant at sunrise, a grown man at noon, and an elderly man at sunset. In later Egyptian times, Ra and Horus combined into Re-Horakhty and commanded both the sky and the Earth.

Ra may have been a pseudo-monotheistic god—that is, the one truly central god of Egypt. Some scholars think that the other Egyptian gods were manifestations or forms of Ra.

Some of Ra's symbols were the bird, a symbol of fire and re-birth; a sun disk; an ankh, representing the life that the sun provides; an obelisk, representing the sun's rays; the pyramids; the bull; and a cobra wrapped around the sun.

The ankh, an Egyptian hieroglyphic symbol for life, looks like a cross with a loop on top, and Egyptian gods were often shown holding one or two ankhs. Although it might be thought of as a staff, it looks like a cross more than it does a rod.

The obelisk, on the other hand, resembles a very tall pencil. The ancient Egyptians placed an obelisk on either side of temple entrances. Symbolizing Ra, an obelisk was thought to be a sun ray in which Ra himself existed.

Ra's establishment as the sun god grew in the Second Dynasty, and by the Fourth, the Egyptians believed that their pharaohs were sons of Ra, or manifestations of the god on Earth. By the Fifth Dynasty, the pharaohs were building solar temples, pyramids, and obelisks in his honor, and he was established as a state deity. Then, by the Eleventh Dynasty, Ra was elevated even further into a more monotheistic version, whereby he had created the world for mankind, and it was men who did bad deeds and caused evil things to happen. He had become a Christian-like god at this point, and some followers thought that Ra would punish them in death if they did evil deeds in life. Yet later, during Egypt's "new kingdom," the Egyptians inscribed their tomb walls with stories of Ra's journey through the underworld, and the mythology around Ra grew to include such things as Ra bringing messages from the living to the dead as he rode his sun boat to the underworld.

So there was definitely a very powerful and well-known Egyptian god called Ra. Given that all pharaohs seemed to carry staffs, it is a reasonable notion that Ra might carry a staff. It so happens that the hieroglyph for the word *god* is "ntr." The hieroglyph looks like a staff bound with cloth that has a handle on top. In fact, the Egyptian title of *pharaoh* is written much the same way: like a staff with a handle on the top.

Staffs were prevalent in ancient Egypt, used by shepherds to tend to their flocks. Staffs were also common religious symbols, connected to Moses, who brought down from Mt. Sinai the Ten Commandments that were put into the Ark of the Covenant.

In the story of the ten plagues, God inflicts Egypt with ten horrors, forcing the pharaoh to free the Israelite slaves. Before inflicting the first plague, God tells Moses to approach the pharaoh and ask for the Israelites' freedom. Moses goes with his brother, Aaron, to see the pharaoh, who refuses to let the Israelites go. So God tells Moses to return to the pharaoh with a warning sign, and when Moses and Aaron return to the pharaoh, the staff turns into a long, writhing serpent.

In those times, because the staff was used to herd flocks, it was considered a symbol of authority. Moses used his staff not only to tend a flock of sheep, he used it to direct the Israelites through the desert. When God told Moses to use his staff to part the Red Sea, he did so and it worked. And Moses' staff was also able to squeeze water from a stone.

When the Israelites rebelled against Moses' announcement that the tribe of Levi would be the priests, God told each of the twelve tribes of Israel to provide a staff. The staff belonging to the tribe that would become the priests would sprout buds overnight, said God. For the tribe of Levi, Aaron provided his staff, and overnight, it bloomed and even bore ripe almonds. In honor of this event, Aaron's staff was kept in the Tabernacle.

The Staff of Ra is not known to exist, as far we can tell, although it's certainly possible that the staff of a pharaoh exists, that Aaron's staff is buried somewhere, and that a staff presumably belonging to Ra might be buried among some Egyptian artifacts. But archaeologists have found no traces of such things; hence we conclude that the Staff of Ra was an excellent device created by the makers of *Raiders of the Lost Ark*, based on biblical and ancient Egyptian references.

Tanis, the Lost City of Ancient Egypt

In Egypt, Indiana Jones and Marion meet up with Indy's friend Sallah, who is digging at the Tanis archaeological site. Sallah tells Indy that the Nazis have hired Rene Belloq, a French grave robber and Indy's rival, to find the Ark of the Covenant. In a busy marketplace, the Nazis kidnap Marion and fake her death. That night, Indy and Sallah take the headpiece to a man who is able to decipher its symbols.

Indy and Sallah realize that the Nazis have miscalculated the resting place of the Ark and are digging in the wrong location. One side of the headpiece provides a partial measurement for the staff's length, but the other side of the headpiece gives a critical correction to that measurement. Without both pieces of information, the digging location is wrong. The Nazis managed to retrieve only the partial measurement from an image of one side of the headpiece that was burned into Major Toht's hand when he tried to retrieve the headpiece from the fire at the Raven Bar.

We've touched on Tanis a bit in previous chapters, and we've talked about lost cities in the South American jungles. Was Tanis also a lost city, albeit one in Egypt? Did the Germans discover Tanis in the 1930s?

First, it should be pointed out that Tanis is a real city in Egypt. Its modern name is San el-Hagar, and, in fact, Tanis is the Greek name of the ancient city called Djanet. It is on the Tanitic branch of the northeastern Nile. During the late Twentieth Dynasty, the ancient Egyptians began construction of Tanis. The Eighteenth through Twentieth Dynasties were called the New Kingdom, and the driving force of the final period of the New Kingdom was Ramses III, the second pharaoh of the Twentieth Dynasty. Ramses III ruled ancient Egypt from 1186 to 1155 BC.

During the Twenty-First Dynasty, the pharaoh Hedjkheperre Setepenre Smendes founded Tanis as the northern capital of Egypt and made his home there. Smendes owned land in lower Egypt, where he seized control from Ramses XI and took hold of the throne. While Smendes controlled lower Egypt, the middle and upper sections of the country were ruled by the high priests of Amun. Smendes' full name meant "Bright Is the Manifestation of Re, Chosen of Re-Amun." It is possible that the Staff of Ra in *Raiders of the Lost Ark* is based on the fact that the founding ruler of Tanis was Smendes, the Manifestation of Ra.

In the Twenty-Second Dynasty, Tanis remained a significant political and commercial city, and it continued to serve as Egypt's main capital.

In 31 BC, Augustus Caesar of the Roman Empire conquered Egypt and stationed military garrisons to keep the peace. During the Roman Period, Tanis did indeed sink into silt, for the most part, and it became a minor village, with people burning the temple limestone to extract the lime.

In AD 6, Tanis was abandoned by its inhabitants when the water of Lake Mazala rose to such levels that it threatened to drown the entire city. The people of Tanis escaped and built a new city nearby, which they called Tennis.

In *Raiders of the Lost Ark*, Tanis is a lost city that the Germans discover during their search for the Ark of the Covenant. According to the movie, the Ark is buried in a secret temple chamber somewhere in Tanis. Was the real Tanis buried by a sandstorm, lost for centuries until the Nazis found it in the 1930s? That is an interesting fiction created by the makers of the film. In fact, the main excavations of Tanis occurred long before the Nazis existed.

Tanis does contain many ruins, and among them are quite a few temples, including a significant temple dedicated to Amun, the primary god of Tanis. With Amun were two other Tanis gods, Mut, who was Amun's consort, and their child, Khonsu. The earliest recorded building in Tanis is from the period 1039–991 BC, in the Twenty-First Dynasty; it is an enormous mud-brick wall at the Temple of Amun. This outer, enclosing temple wall is approximately 16 yards thick, and it is approximately 470 by 400 yards wide. Within this outer wall is yet another mud-brick wall.

The stones of the temples were carried by the Egyptians from the town of Qantir. In the late Twenty-First Dynasty and the Twenty-Second Dynasty, the Egyptians added more buildings to the main Temple of Amun, including a small temple dedicated to Mut and Khonsu on the southwest side of the main site.

Napoleon Bonaparte surveyed Tanis in the late 1700s, and the first true exploration of Egypt occurred when Bonaparte invaded the country. He sent scholars along with his army; the scholars conducted surveys of sites all over Egypt. These early Egyptologists recorded what they found, but they did not plunder artifacts as did the treasure hunters who followed them.

In the early 1800s, explorers were more interested in collecting and selling Egyptian antiquities than in simply recording their existence. Most of the excavations during this time, however, involved statues. Jean-Jacques Rifaud found two large pink granite sphinxes and brought them to Paris. These statues are now in the Louvre. Statues from Tanis also ended up in Berlin and Saint Petersburg, and in the early 1800s, Henry Salt and Bernardino Drovetti unearthed eleven more statues, shipping them to the Louvre, Berlin, and Alexandria.

The first major excavation of Tanis was performed by French Egyptologist August Mariette from 1860 through 1880. Mariette is often credited as the founder of modern archaeological excavations in Egypt. In addition to his major contributions to Egyptology, Mariette also conceived the plot for what became the opera *Aida* by Verdi.

At any rate, between 1860 and 1864, Mariette discovered the Four Hundred Year Stela, an enormous granite monumental plaque that had been built in the eastern Delta by Ramses II in honor of Seti I, his father. Mariette also dug up many royal statues dating back to the Middle Kingdom: the Eleventh and Twelfth Dynasties between 1986 and 1759 BC.

Later excavations were done in 1883–1886 by Flinders Petrie and from 1921 through 1951 by French Egyptologist Pierre Montet. Petrie found a Roman-era papyrus, and he made detailed drawings of the temple precinct. He also excavated trenches and recorded inscriptions on the walls that he uncovered.

The treasures unearthed by Pierre Montet were magnificent. Ten years into his digging in Tanis, in 1939, Montet found the tomb of King Osorkon II (874–850 BC). The site had several rooms, all of which had been plundered. Despite the earlier robberies, the tomb still contained many amazing artifacts, among them the quartzite sarcophagus of Takelot II (850–825 BC), who was Osorkon's son, as well as various bejeweled objects.

Montet completed his excavation of the main tomb, and then, to his surprise and delight, he located another tomb. This one had remained undisturbed by robbers. It was the tomb of Shoshenq II, with a solid-silver coffin containing a solid-gold face mask and beautiful gold jewelry. Perhaps most amazing, Shoshenq II was a king whom Egyptologists did not yet know about until Montet's find. The tomb of Shoshenq II also contained the sarcophagus of Amenemope (993–984 BC). Altogether, Montet discovered six royal tombs, all of them subterranean and made from mud bricks and stone blocks that were in large part inscribed.

The importance of Montet's finds remains impressive today. The artifacts unearthed by Montet are the most significant source of knowledge about royal funerary objects during what is known as the Third Intermediate Period of Egypt's history. This is a very long period of time in Egypt, ranging from the death of Ramses XI in 1070 BC to the founding of the Twenty-Sixth Dynasty in 664 BC by Psamtik I.

The Tanis treasures are stored in a museum in Cairo and include items such as

- The solid-gold funerary mask of Psusennes I
- The solid-gold funerary mask of Shoshenq II
- A gold pectoral with inlays of multicolored stones and glass, from the neck of Shoshenq II's mummy
- The solid-gold sandals worn by Shoshenq II's mummy
- Gold, scarab, and lapis lazuli jewelry, with embedded green and red faience (ceramics)
- Seven pairs of hinged bracelets featuring the wadjet eye (the symbol of the Eye of Horus) and made from gold, lapis lazuli,

and carnelian, with embedded green and red faience, all from
the mummy of Shoshenq II

- The solid-gold funerary mask of General Wendebauendjed
- A solid-gold bracelet inlaid with a scarab
- A solid-gold pectoral with a solar motif and embedded lapis
  lazuli

Lapis lazuli, a gemstone highly prized by the Egyptian pharaohs, was fashioned into amulets and ornaments such as scarabs.
The ancient Egyptians worshipped the *Scarabaeus sacer*, a dung beetle, as the embodiment of the god Kheptri. Scarabs were amulets,
or good luck charms, made in the shape of these beetles. Lapis lazuli is a deep-blue, opaque jewel, and the finest lapis lazuli is a very
intense blue with light flecks of golden pyrite, commonly known as
fool's gold. In the Egyptian Book of the Dead, lapis lazuli in the
shape of an eye set in gold is said to be an amulet with enormous
power. In fact, it was considered to be so important that Egyptians
made offerings to the symbolic eye on the last day of every month.
In addition, Egyptian women wore lapis lazuli dust as eye shadow.
The pharaohs got their lapis lazuli from mines at Shortugai on the
Oxus River in northern Afghanistan. These mines are still operating today.

Today's Tanis is a desolate place, covered in silt and surrounded
by an open plain. During the last two millennia, the inhabitants of
the area have consisted mainly of wild boar and transient Bedouins.
The site of Tanis today contains debris and fallen statues and columns, as well as reused obelisks of Ramses II and temple blocks
from ancient times. The enormous enclosing walls are mostly gone.
People can enter the site from various routes. The typical way that
people enter the site, however, is through the entrance to the ruins
of Sheshonq III. Two deep wells that once showed the height of
the Nile water are in the middle of the Temple of Amun.

In the northern corner of the Tanis site, south of the wall of
Ramses II in the Temple of Amun, is the ancient Sacred Lake. This
lake is rectangular and lined with stone, and it originally had stair-
ways leading down into the water. Tuthmosis III dug the Sacred

Lake in accordance with what is thought to be a common practice of including sacred lakes in most temple sites. The Sacred Lake at the Temple of Amun was filled with groundwater and used for ritual cleansings. To the ancient Egyptians, the Sacred Lake represented the primeval waters that spawned life.

In the northern corner of the Sacred Lake was a huge granite statue of a scarab on a cylindrical pedestal. There was a stela on the flattened front face of the pedestal, and on this stela, a king was depicted in a kneeling position. The statue was destroyed during the course of many centuries, like most buildings in this area.

On the south side of the Sacred Lake is a stone tunnel. In ancient times, the Egyptians released the geese of the god Amun through the tunnel into the lake from special yards that were built nearby to contain the waterfowl. The priests had their homes around the Sacred Lake; ruins of these homes still remain on the eastern bank.

The Well of Souls, Temples, and Map Rooms

After making his way to Tanis, Indiana eventually infiltrates the Nazi digging site and starts to look for the Ark in its headpiece-designated location. The Nazis, with their hundreds of Arab workers, are everywhere with equipment and excavation platforms. They are frantically searching for the Well of Souls, which, according to legend, is where they will find the Ark of the Covenant.

Indy and Sallah see a round hole in a mound of dirt, where the sun will be positioned exactly overhead at nine in the morning. Using a long rope, Indy drops through the hole into a room, the floor of which is elaborately decorated with a detailed map of ancient Tanis. Indy deciphers the hieroglyphics that are engraved in the stone floor, and he figures out where to put the Staff of Ra. When he positions the staff, suddenly sunlight pours through the hole onto the map, moving along the floor, then hitting the top of the headpiece at a particular angle. A beam of light shoots from the headpiece and focuses on a building on the floor map of ancient Tanis.

This is the location of the Well of Souls, the resting place of the Ark of the Covenant. It is nowhere near the location where the Nazis are digging.

We've already discussed the ruins of Tanis and the Amun Temple in some detail. Was there really a Well of Souls, and was it underground, requiring that Indiana Jones break through a door on its roof? Or is the film's Well of Souls a reference to the two deep wells that once measured the height of the Nile water in the middle of the Amun Temple?

While there is no reference to the two real wells in the movie, the reference to the Well of Souls is somewhat valid, though misplaced. In other words, there is a Well of Souls, but it is not located in Tanis, Egypt.

The real Well of Souls is a cave that is beneath the Sakhrah in the Dome of the Rock. The Dome of the Rock is a famous landmark in Jerusalem that was built between AD 687 and 691. The Sakhrah is the center stone in the Dome of the Rock, which is also known as the Pierced Stone because it has a tiny well-shaped hole that is on the roof of the cave. A staircase leads through a gap between the Sakhrah and the surrounding bedrock down into the cave, which is about the size (width and length) of the overhead Sakhrah. The ceiling curves downward on all sides, and where the steps are located—in the southern part of the cave—are man-made walls that support the roof of the staircase.

Muslims believe that Muhammad rose to God in heaven with the angel Gabriel via the Sakhrah. Once in heaven, Muhammad talked to Moses and received the Islamic prayers that are now mandatory in that religion. Then he returned to Earth.

For Jewish people, the Sakhrah was where Abraham was willing to sacrifice his son Isaac at God's command. Long ago, the Sakhrah was inside the First Temple, and the Ark of the Covenant rested upon it. During the time of the Second Temple, the high priest sprinkled the blood of sacrifices on the Sakhrah and placed offerings to God on it.

Christians believe that the Dome of the Rock was constructed in the same spot where Constantine's mother built the Church of St. Cyrus and St. John, which was later called the Church of Holy Wisdom.

According to the teachings of Islam, the Last Judgment will occur at the Sakhrah, and the souls of the dead are now gathering in the cave, in the Well of Souls, waiting for the judgment day. A great wailing is said to echo from the Well of Souls, which is made by the voices of the dead calling out upon the Rivers of Paradise that flow over the Abyss of Chaos. Another, more objective, source of the wailing noises might be the type of resonance that is heard inside seashells.

As in *Raiders of the Lost Ark*, legends claim that the Well of Souls was the hiding location of the long-lost Ark of the Covenant. Whether this is true, as mentioned earlier, is unknown, and it's highly unlikely that government and religious authorities will let archaeologists dig for the Ark beneath the Temple Mount.

So, yes, the Well of Souls is real, but no, it isn't in Tanis, Egypt. It's a cave beneath the Dome of the Rock in Jerusalem. And there is no known map on the floor that shows precisely where to find the Ark of the Covenant.

Speaking of the map, how likely was it that a beam of light hit the headpiece of the Staff of Ra and illuminated a building on the map? This one is not as unlikely as you might think. Many tombs around the world were constructed so that the sun would send shafts of light to shine on specific locations.

For example, the Neolithic tomb Newgrange in Ireland is somewhat like the map room in *Raiders of the Lost Ark*. This tomb was built in approximately 3200 BC and covers more than an acre. According to mythology, the mound over the tomb was reputedly the site of a fairy mound, the home of Oenghus, the god of love. Richly decorated stones surround the structure, and a twenty-foot-long inner tunnel leads to a main chamber. Side chambers are situated off the tunnel. A slit over the roof of the tunnel's entrance lets a shaft of sunlight in, which beams down the tunnel and brightly illuminates it. This happens only during a very specific time of year, at approximately the time of the winter solstice, from December 19 to 23.

Another example might be the Temple of Serapis. It contained a narrow window on one side, set into the wall so that every morning at dawn, the sun shone precisely through the window. A ray from the sun traveled across the floor to the statue of Serapis (a Hellenistic-Egyptian god) and up to the statue's lips. It thus seemed that every morning, the sun rose and kissed the statue. It is thought that the walls were covered in gold with silver and bronze embellishments, and that these precious metals somehow magnified and held the image of the sun kissing the statue for a long period of time. The Temple of Serapis included marble columns and platforms that rose up a hundred stairs, then down again. The statue itself was huge, with its right hand touching one wall and its left hand on the opposite wall. The temple stood intact for six centuries but was destroyed in AD 391 by Theophilus, the bishop of Alexandria.

In *Raiders of the Lost Ark*, the Staff of Ra catches the sun's rays and shoots them in another direction. This is slightly different from temples with slits and windows that attract sunlight once a year or once each day. There is actually a temple called the Shrine of Remembrance that features the very same mechanism. It is located in Melbourne, Australia.

The Shrine of Remembrance was built in honor of soldiers who fought in World War I. The architect, Philip B. Hudson, constructed the memorial so that for at least five thousand years, at exactly 11 a.m. on November 11, the anniversary of the end of war, a ray of light will streak through a hole, pass over the Stone of Remembrance, and illuminate the word *love* in the inscription, "Greater love hath no man." The Stone of Remembrance is made of marble and sits beneath the pavement where people cannot reach and touch it.

In 1971, the state of Victoria, where the shrine is located, adopted daylight saving time for the summer. In this part of the world, summer happens to include the month of November, so 11 a.m. on November 11 became noon, an hour later than the sun was supposed to hit the word *love*.

Using a lighting method similar to the one for the Staff of Ra, the problem with the Shrine of Remembrance was fixed by installing a mirror to deflect the sun into the hole at 11 a.m.

Deadly Snakes

As night falls and lightning flashes, Indiana and a small crew start to dig at the location specified by the map. They unearth a door leading to a subterranean chamber. It is the Well of Souls, and it is filled with deadly snakes.

Indy and his men, knowing that snakes hate fire, throw torches into the underground chamber to clear the snakes. As Indy lowers himself into the Well of Souls, he falls and lands directly in front of a cobra. After slowly recovering, he sprays kerosene from canisters around the chamber and ignites the snakes with a torch.

At the far end of the Well of Souls is a stone altar holding a heavy stone chest that contains the Ark of the Covenant. Indy and Sallah heave the top of the stone chest up and slide poles through the rings on the sides of the chest. They hoist the Ark out. It is solid gold and has two gold angels facing each other on the top. As the sun rises, Indy's men use ropes to pull the Ark out of the Well of Souls.

But then Belloq and the Nazis steal the Ark, toss Marion into the cavern, and seal the cavern roof, entombing Indiana Jones and Marion in the Well of Souls. Their torches are beginning to expire, and snakes are everywhere.

Indiana Jones is not alone in having a fear of snakes. Many people do indeed fear them as much as he does. Experts think that our fear of snakes is rooted in our earliest history, when mammals had to survive in a deadly environment filled with killer reptiles. According to a *National Geographic* article, psychologist Arne Öhman of Stockholm, Sweden, coauthored a 2001 study on the subject that was published in the *Journal of Experimental Psychology: General.*[11] Öhman concluded that to survive in such a harsh early environment, we developed the ability to perceive and focus on snakes and spiders, which we saw as life threatening, and to respond instantly with terror, which enabled us to protect ourselves and our loved ones. In the same *National Geographic* article, Joseph LeDoux, a professor of neural science and psychology at New York University, agreed with Öhman's conclusions, adding that "there are certain stimuli that are pre-wired in the brain because they have been perennially dangerous to our ancestors."

There is a term for the fear of snakes. It is sometimes called ophidiophobia, although it is more commonly known as herpetophobia, which means fear of reptiles. This is one of our most common phobias.

It is also possible that we are afraid of snakes for other reasons that do not stretch back to the dawn of time. After all, major religions such as Christianity view the snake as the physical embodiment of evil on Earth. It was a snake that seduced Eve into eating the apple.

The most likely reason that people are afraid of snakes is simple. Snakes look terrifying and dangerous, and they do bite with fangs that inject poisonous venom. Some snake bites are fatal to humans. These facts are enough to instill fear in anyone who is unskilled in handling snakes.

Snakes have been slithering and biting their way around the planet for more than 150 million years. Australia is host to seventeen types of lethal snakes. Throughout the world, there are probably between 1 and 2 million dangerous snake bites reported each year, and some estimates place deaths from snake bites at approximately 50,000 a year.

In Africa, deadly snakes include the Egyptian cobras, as well as saw-scaled vipers and puff adders. Asia has the common cobra and the Russell's viper, and South America is home to the deadly anaconda.

Let's focus on the deadly Egyptian snakes, such as the cobras and asps that Indiana Jones encounters in *Raiders of the Lost Ark*. The Egyptian cobra has another name: the Egyptian asp. This is the most common cobra in Africa, and it causes more human deaths on that continent than any other type of snake. This snake was used by Cleopatra to end her life, and it was the pharaohs' symbol of sovereignty.

The Egyptian cobra has a large, depressed head with a wide snout, and its eyes are big with round pupils. Its neck can dilate or widen into a hood shape, which spans six to seven inches. Its body is thick, cylindrical, and long. The cobra's average length is from three to six feet, although some grow to nine feet.

The Egyptian cobra is divided into several species, such as the *Naja haje haje*, which is found south of the Sahara desert; the *Naja haje legionis*, in Morocco; and the *Naja haje Arabic*, in southwest Arabia. The *Naja haje haje* has the largest geographic range and lives in many habitats. It is found in many countries, including   Algeria, Cameroon, Egypt, Ethiopia, Guinea, Ivory Coast, Kenya, Libya, Morocco, Niger, Nigeria, Saudi Arabia,

Sierra Leone, Somalia, Sudan, Tanzania, Uganda, and Yemen, among others.

These cobras live almost anywhere, so they could indeed be found in the underground cave that Indiana Jones is trapped in. Egyptian cobras live in the grasslands, on hills with some vegetation on them, in agricultural areas, in oases, in steppes, on dry savannahs, and in deserts with sparse vegetation. They also reside in houses and villages, and they swim in the Mediterranean Sea.

When the Egyptian cobra perceives that it might be in danger, it raises its hooded head. Its bite kills quickly, and its venom is more potent than the venom of any other cobra except for the cape cobra of southern Africa. But the Egyptian cobra is much larger than the cape cobra, and it injects far more venom in each bite.

Other horrific snakes with expanding hoods that are found in Africa are the black-necked cobra, which spits venom at its victim's eyes from as far away as seven feet, and the black mamba, whose bite is always fatal unless antivenoms are immediately administered. The black mamba is huge, reaching a length of fourteen feet, and it rears up before striking a large animal or a person. In addition, the gaboon viper of Africa is extremely dangerous because its fatal bite is usually confused with an insect sting, and people do not seek medical attention quickly enough. Many could be saved by the antivenom if only they realized they were bitten.

The venom injected by killer snakes is a mixture of toxins and proteins that immobilize and digest prey. This prey could consist of smaller snakes, as well as mice, birds, and frogs. Snake venom attacks the victim's heart, lungs, muscles, and/or red blood cells. Hemotoxic venom attacks the blood vessels and induces hemorrhaging. Neurotoxic venom paralyzes the victim's heart and lungs. And myotoxic venom attacks the muscles, inflicting enormous pain. Some snakes have multiple types of venoms and with one bite can induce multiple effects.

Cobras, such as the ones encountered by Indiana Jones, have extremely potent neurotoxic venom. Coral snakes also have potent neurotoxic venom, and in this case, the venom liquefies the flesh of the victim. A bite from a cottonmouth snake can cause a fatal amount of hemorrhaging fairly quickly.

In ancient Egypt, snakes were everywhere and could easily have been in underground caves. They lived in the desert, by the Nile River, and in houses and walls. Some were poisonous and in fact deadly to both livestock and humans. For their own safety, the ancient Egyptians had to keep snakes at a distance. They viewed the reptiles as protecting the king while also being demons from the underworld.

The horned viper, also known as the sand viper, was greatly feared in ancient Egypt. This type of snake, when it attacks, rasps its coils together, repeatedly making a long *fy* sound. (This *fy* sound is heard in the Egyptian word for "viper.") And then the snake leaps at its prey.

Various medicinal and religious texts refer to the threat of snakes. The Pyramid Texts, containing ancient Egyptian spells, refer often to the deadly attributes of snakes. The ancient Egyptians feared the snake god Apophis, or Apep, as the enemy of Ma'at, the god of all order in the universe, without whom the universe would collapse. When Ramses II was pharaoh, Apophis was a symbol of great evil. Certain religious rituals consisted of torturing images of the serpent god made from papyrus and wax, to represent the triumph of Re and Ma'at over chaos.

Do snakes really hate fire? Will a torch really protect someone from snakes? In actuality, snakes avoid bright lights, so taking a torch into the darkness to protect yourself from snakes is a good idea. Indiana Jones's use of the torch was wise and is one of the most common pieces of advice given to people who venture into dark areas that are infested with snakes.

Pyramid Building

When Belloq and the Nazis entomb Indiana Jones and Marion in the Well of Souls, Indy climbs up a pillar and pushes against the top of the chamber. The pillar starts to rock and quickly topples, destroying the chamber wall.

It's actually not all that easy to destroy a pyramid structure. These architectural wonders have been standing tall throughout time.

The Great Pyramid on the plateau of Giza is one of the original Seven Wonders of the Ancient World, and it was constructed from 2589 to 2566 BC.

Most people believe that slaves built the Egyptian pyramids, although some people conjecture that skilled workers receiving salaries were employed to create the structures. What we know for sure is that the earliest pyramids were made entirely of stone. The outer casing was of a high-quality limestone quarried at Tura, which is near Cairo, and the main pyramid body was built from limestone that the Egyptians quarried locally. The porticulis (a gate), and the roofs and the walls of the burial chambers were often made from granite that was transported from Aswan.

The earliest pyramids had a less stable structure than the later versions had, although they were still extremely strong and not so easily destroyed by one man. In these earliest structures, the Egyptians placed layers of stone that sloped inward, but later they learned that the pyramids were stronger if the stones were stacked horizontally, one on top of another. Some pyramids even used both techniques, as the Egyptians made the transition from sloped to horizontal stones.

During the Middle Kingdom, the Egyptians were building pyramids from mud bricks encased in limestone. These pyramids did not survive the centuries as well as the earlier monuments, which were carefully constructed of massive stones in stable configurations.

The gaps between stones were filled with a special, very strong mortar made from gypsum, rubble, and wood. The ancient Egyptians first dehydrated the gypsum-rubble mortar by heating it over wood fires. In fact, much of Egypt's natural wood-tree resources were destroyed to build the pyramids.

The Egyptians were very exacting in the construction of the pyramids, which is why they are extremely strong and have stood the test of time, despite their incredible height. Workers used saws, drills, and copper chisels to cut limestone, and they used abrasives such as quartzite sand, along with drilling, sawing, and pounding with dolerite, to cut granite. They transported the stone blocks on sledges that they lubricated with water. Foundations were leveled, as were the tiers of the pyramid.

When a pyramid was finished, it might weigh something like six million tons or more. The Great Pyramid is taller than the Statue of Liberty and contains more than two million stone blocks, each weighing approximately 2.5 tons. At its base, the Great Pyramid is wider than ten football fields. Clearly, one man cannot destroy the walls of one of these ancient structures.

Seaplanes and Luftwaffe Airplanes

When Indiana Jones is en route to Nepal, he boards a Pan American clipper seaplane and flies over the Pacific Ocean from San Francisco to the Himalayas. There, he intends to find Professor Ravenwood, who has the Staff of Ra headpiece. The seaplane has several rows of seats.

Much later, after Indiana surfaces from the Well of Souls, he sees that the Nazis are preparing to take the Ark away on a Luftwaffe airplane. Indy blows up the plane, and the Nazis put the Ark on a truck.

Did such a Pan American clipper seaplane exist in the 1930s? It sure did, but it was expensive. One ticket across the Pacific Ocean aboard a Pan American clipper cost the equivalent of $10,000 in today's currency.

The founder of Pan Am was Juan Trippe. He wanted to create a style of flying that would rival travelers' experiences aboard luxury ocean liners, so he built his 1930s clippers with what he hoped were equal amenities: fine food, fine drink, and lavish service. There were three types of Pan American clipper seaplanes: the Sikorsky S-42, the Martin M-130, and the Boeing 314.

The name *clipper* was chosen because Juan Trippe's family had amassed its fortune centuries earlier by sailing clipper ships. The 1930s clipper planes were referred to as flying boats, and they could take off from and land on water. There were few runways in those days, and they were very expensive to build, so using water as a runway made a lot of sense. Like the Trippe family's clipper ships from long ago, the Pan Am clipper seaplanes also crossed the oceans.

The first sea clipper introduced by Pan Am was the Sikorsky S-42 in 1934, which flew from Miami to Buenos Aires and held thirty-two passengers. Its fuel tanks barely made the twelve-hundred-mile journey.

In late 1935, Pan Am introduced the Martin M-130, nicknamed the China Clipper. It flew across the Pacific Ocean, and in 1936, it took passengers on an eight-thousand-mile, weeklong trip to Hong Kong. Each M-130 could handle forty-six passengers.

Long after Indiana Jones needed a flight to Nepal, Pan Am introduced its third clipper seaplane, the Boeing 314, in 1939. It carried seventy-four passengers, who resided in luxury cabins that included dressing rooms and staterooms, full-course meals, and separate restrooms for men and women. The Boeing 314 took a Pacific route in 1939, but later that year, it also flew over the Atlantic Ocean.

The Luftwaffe was the German air force, formed in May 1935 after the Nazis passed the Law for the Reconstruction of the National Defense Forces. The law returned into existence the elements of war—an army, a navy, and an air force—that had been banned in Germany since the end of World War I.

After World War I ended and the Treaty of Versailles was signed in 1919, the Germans maintained a small defensive army known as the Reichswehr. It was controlled by the Allies, who hoped to keep Germany from instigating future military aggression. The Reichswehr had about a hundred thousand men in service, for both land defense and a small navy. But the Reichswehr did not include an air force.

Then in 1933, sadly for the world, the Third Reich came to power with the National Socialist German Workers Party, and in 1935, the Germans renounced the Treaty of Versailles. They transformed the Reichswehr into the Wehrmacht, which included an army, the Heer; a navy, the Kriegsmarine; and a new air force called the Luftwaffe.

The Luftwaffe consisted of many thousands of airplanes of all types and even included naval vessels. Between 1939 and 1945, more than 3.4 million Germans served in the Luftwaffe, and of the 7,361 men who received the highest German combat award during World War II, the Knight's Cross, 1,785 (or 24 percent) served in the Luftwaffe.

As for the specific types of planes in the Luftwaffe, they were numerous: Arado Ar 68, 96, 196, 232, 234, and 240; Blohm und Voss BC 138 and 222; DFS 230; Dornier Do 17, 18, 24, 215, 217, and 335; Fieseler Fi 156; Focke-Wulf Fw 189, 190, and 200; Focke-Wulf Ta 152 and 154; Gotha Go 242 and 244; Heinkel He 45, 46, 59, 60, 111, 114, 115, 162, 177, and 219; Heinkel Hs 123, 126, and 129; Junkers Ju 52, 86, 7, 88, 90, 188, 252, 290, and 388; Messerschmitt Bf 108, 109, and 110; Messerschmitt Me 163, 210, 262, 321, 323, and 410; and Siebel Si 204.[12]

Cliff Roads Near Cairo

When the Nazis put the crated Ark of the Covenant in the back of a truck, Indiana Jones hops onto a white Arabian stallion to chase the truck in order to recover the Ark. Belloq is riding in a staff car in front of the truck. From the top of a cliff near Cairo, Indiana sees the car and the truck far beneath him. He charges after them and gallops down the steep hillside, finally catching up to the Nazis. Indy leaps from the horse into the truck, eventually throws the driver onto the road, and grabs the steering wheel.

Yes, there are cliffs outside Cairo, and this scene could actually take place. Obtaining a white Arabian stallion might be tricky, but nothing is too difficult for Indiana Jones.

Egypt is bounded in the south by Sudan, in the west by Libya, in the east by Israel, and in the north by the Mediterranean Sea. More than 90 percent of Egypt is desert, with less than 10 percent of the country settled or used for agriculture and livestock.

Cairo is the capital of Egypt, with a population of approximately 16.1 million people. It has the largest population of any metropolitan area in Africa, and it is the seventeenth most heavily populated metropolitan area in the world. Cairo is fairly close to the Mediterranean Sea and lies on the banks of the Nile River. The city is immediately south of where the Nile River splits into two branches into the Nile Delta region.

The Arabian Desert stretches to the Gulf of Suez and to the Red Sea, and much of it is on a plateau that approaches an elevation of 2,000 feet in the east. This plateau is punctuated by tall, jagged peaks that reach 7,000 feet along the coast of the Red Sea. The mountains on the Sinai Peninsula include Jabal Katrinah, which is 8,668 feet high. The Nile River flows north from Sudan through Egypt for 960 miles to the Mediterranean Sea. For its entire length, from the southern border of Egypt all the way to Cairo, the Nile River flows through a narrow, cliff-lined valley.

Nazi U-Boats

Indiana Jones and Marion, in possession of the Ark of the Covenant, intend to sail from Cairo to England on the Bantu Wind *steamer. The morning after they set sail, the steamer becomes eerily quiet. Indiana cocks his gun and goes onto the deck to try to discover what is wrong.*

Germans in a U-boat that has been following the ship carrying the Ark have taken hold of the steamer, and a party of Nazis has come aboard. The Nazis kidnap Marion again, and again steal the Ark. They head into the sea in their submarine, with Indiana Jones clinging to the top of it.

The German U-boat did indeed exist, even before the 1930s, although from the end of World War I to the beginning of World War II, U-boat activity was greatly diminished. It is possible, albeit strange, that Indiana Jones encountered a U-boat during the time of *Raiders of the Lost Ark*.

The German word for the submarine is *Unterseeboot* (undersea boat), or U-boat for short. The U-boats were German military submarines that operated in World War I and World War II. In both wars, U-boats attacked merchant convoys that were hauling supplies to Europe from the United States and Canada.

When World War I broke out, Germany had twenty-nine U-boats, and within weeks, the submarines had destroyed fifty British ships. Then in 1915, a U-boat sank a passenger ship, the RMS *Lusitania*, with one torpedo. Aboard the ship were 1,198 people, all killed. The Allies were stunned that the Germans would attack an unarmed passenger ship. The manifest of the *Lusitania* indicated that the ship was carrying some nonexplosive military cargo, but Germany had signed an agreement before the war that prohibited the country from sinking any passenger ship, even one with a small amount of nonexplosive military cargo onboard.

Between October 1916 and January 1917, German U-boats sank a million and a half tons of shipped goods, and in late January, the Germans told the world that the U-boats would now be launched into outright warfare of an unrestricted nature. Then in March, German U-boats sank three American merchant vessels. At this point, the United States declared war on Germany. Eventually, the war ended in November 1918. At the Treaty of Versailles in 1919, Germany was forbidden to build any more U-boats, and its use of current boats was limited.

Under the guise of research, the Germans started to build more U-boats in hidden locations. When World War II broke out, the Germans were ready with a small fleet of new U-boats. In fact, during that war, Germany's fleet of submarines exceeded in size the fleet of any other country.

The World War II U-boat was basically a launching pad for torpedoes, which either exploded as soon as they hit something solid or exploded when they sensed large metal objects. Using the second kind of explosion, the Germans would launch the torpedo so that it barely missed hitting an Allied ship. The near miss would create a gas bubble that was strong enough to break the Allied ship into pieces.

As for Indy clinging to the top of the Nazi submarine, it must be pointed out that the Nazi U-boats were in fact submersibles, not true submarines, and they did most of their fighting on the surface as if they were torpedo boats. They submerged only to avoid enemy ships, to escape the weather, and to make a rare daylight attack by periscope. They usually traveled on the surface, too, where they were much faster and more maneuverable. Because the Nazis wanted to get the Ark to its destination as fast as possible, the U-boat would have traveled above water all the way to the hidden island, enabling Indy to safely hitch a ride.

Secret German Submarine Bases

The Nazis, with Indiana Jones stealthily clinging to the top of their U-boat, make their way to a remote island in the Aegean Sea, southeast of Greece.

Did the Nazis have secret German submarine bases in the 1930s as they did in *Raiders of the Lost Ark*?

During World War II, warfare was intense in the waters around the Aegean islands, with the people on the islands nearly starving to death. Nazi submarine bases did exist in Greece, and these were bombed by the British, to the extreme joy of the starving population. In October 1943, the Germans seized Kos, the only Allied air base in the Aegean Sea. But in the 1930s, was there a secret island somewhere that had a German submarine base? It is unlikely.

Arcane Jewish Ceremonies and Sacred Breastplates

On the hidden Nazi submarine base in the Aegean Sea, Belloq begins a mysterious Jewish ceremony that will uncover the secrets of the Ark. Indiana threatens to destroy the Ark with a rocket launcher unless the Nazis release Marion. But Indy's threat doesn't work, and both he and Marion end up being tied to a stake. Belloq and the Nazis are able to open the Ark. It emits a strange beam of light, and frightening spirits appear. Indy orders Marion to close her eyes, but the Nazis stare at the Ark in awe, and they are destroyed by the Ark's incredible power. Thunder booms, and the Ark closes once again.

Just how strange is this scene? Are there any Jewish biblical or mythical references to cryptic Ark ceremonies and sacred breastplates? You may be surprised to learn that there is an element of truth to all of this.

In the Hebrew Bible, God instructs the Jews to make their priestly garments, particularly those for the high priest, with special care. The high priest is to wear a breastplate, an ephod, a robe, a tunic, a headdress, and a sash. The tunic was a fringed garment worn beneath the robe and the ephod. The headdress had a frontlet of pure gold, engraved with the words *Holy to the Lord.*

The ephod was part of a religious device that was used to foretell the future. The ephod was made from linen and was rather small. In a biblical context, the ephod comes across as a pocket. The Book of Exodus specifically indicates that the sacred breastplate was worn on the ephod, and that the ephod was embroidered with threads of gold, purple, red, and blue. The ephod included two shoulder straps with gold rings attaching them to gold chains, which in turn were connected to the sacred breastplate. On the shoulder straps were two jewels made from lazuli, which had the names of the twelve tribes engraved on them. Six tribal names were on one stone, and six on the other. According to biblical sources, one stone was engraved with Reuben, Simeon, Levi, Judah, Dan, and Naphtali. The other stone was engraved with Gad, Asher, Isaachar, Zebulon, Joseph, and Benjamin. In this way, each jewel had twenty-three Hebrew letters on it.

Used with the Urim and Thummin, which we'll talk about in a minute, the ephod served as an oracle. For example, in the Book of Samuel, when David wants to ask God for advice, he requests that a priest give him the ephod.

The breastplate contained the Urim and Thummim, which were components of the oracle. This breastplate was created in the same colors as the ephod and was made from linen. It was square and doubled. Into this elaborate piece of cloth, the Jewish people were commanded to put four rows of precious jewels, with the total number of jewels equaling twelve, to stand for the twelve tribes. The first row contained a carnelian, a chrysolite, and an emerald. The second row contained a turquoise, a sapphire, and an amethyst. The third row contained a jacinth, an agate, and a crystal. And the fourth row contained a beryl, a lapis lazuli, and a jasper. These jewels were mounted in gold on the breastplate.

The instructions continue, with elaborate descriptions of how gold braids, loops, and chains are created and used to hold the breastplate to the ephod. The robe is also described in elaborate detail; in short, it was blue with a hem of blue, purple, and crimson intertwined with gold bells and pomegranates.

Aaron was instructed to wear the breastplate whenever he entered the sanctuary. In this way, he would always remember all twelve tribes when he prayed to God. And inside the breastplate were the Urim and Thummim, close to Aaron's heart when he asked God for advice and indications of what might come. Should Aaron not wear the breastplate, ephod, robe, tunic, headdress, and sash while in the sanctuary, he risked death, for the Hebrew Bible specifically said that he must wear those garments and accoutrements "that he may not die."

In *Raiders of the Lost Ark*, Belloq, acting as the high priest, is clearly wearing attire that corresponds somewhat to the descriptions in the Hebrew Bible. He is definitely wearing the breastplate, and from that fact alone, we can assume that he is wearing the Urim and Thummim.

Biblical evidence pertaining to the Urim and Thummim is slim and unclear. The text introduces the devices as if the readers already know exactly what they are and how they are to be used. Because Moses had direct access to God, he did not require the Urim and Thummim, so in the Five Books of Moses (Torah), the devices are never used. Yet Joshua is ordered that if he ever needs to ask God for advice, he should ask the priest Eleazar to make the inquiries on his behalf, using the Urim. In Numbers, Joshua is told that this is the method he should use to decide whether to send the Israelites into war or not. The ephod was akin to a box that contained the oracle, and the oracle itself was the combination of the Urim and Thummim, although, in some cases, as with Joshua, only the Urim sufficed in order to use the oracular power. When David was fleeing from Saul, he asked the priest Abiathar to bring the ephod to him. Then David asked God through the ephod whether Saul would "come down" and whether "the men of Keilah [would] surrender." God answered yes to both questions. There is no record of God ever answering no to questions asked through the Urim and Thummim oracle. The lack of a "yes" indicated a "no."

When Saul placed a curse on anyone who ate during a battle and Jonathan, who was unaware of the curse, ate some honey, Saul asked the Urim and Thummim to tell him who ate the honey. The oracle pointed to Jonathan.

"Urim and Thummim" are sometimes translated as "lights and perfections." According to ancient stories, when a medallion with God's holy name was slipped into an opening under the oracular devices, the high priest's breastplate glowed and then transmitted messages from God.

As for secret Ark ceremonies, these are unknown. It's possible that the strange beams of light refer to the glow of the breastplate. And maybe long ago, in the time of the Bible, the high priest could use the power of the Urim and Thummim to evoke ghostly characters and thunder, but there is no evidence or mention of these things. It is true, as mentioned earlier in our section about the Ark of the Covenant, that only the high priest, who wore the Urim and Thummim and the breastplate, and so on, could open the Ark. The Ark itself is said to have shot two jets of fire into the path preceding the Israelites when they wandered through the desert. The fire destroyed thorns, scorpions, and snakes. So during the opening of the Ark in *Raiders*, it's conceivable that the Ark shot out some fire, but as for the ghostly characters and the thunder, the roots of these *Raider* ideas are much harder to explain.

The Ark: A Telephone Link to God

In a conversion with Indiana Jones, Belloq claims that the Ark is "a transmitter, a radio for speaking to God."

Is the Ark a telephone link to God? Possibly, because God was said to be in the clouds between the two cherubs on the Ark. God was intimately associated with the Ark, so in theory the Ark could be used as a device to talk to God in some way. The strong implication is that the Ark, when used in conjunction with the Urim and Thummim, which were oracular devices, enabled the high priest to somehow communicate with God.

Top-Secret U.S. Archives

When Indiana Jones and Marion return to the United States, two Army Intelligence officers tell Indy that the Ark is being stored somewhere safe, where experts are studying it. The last thing we see is a crate containing the Ark being wheeled in a huge warehouse that contains similar crates.

Perhaps the most famous example of secret U.S. government warehouses is Area 51, an air force base in southern Nevada. Supposedly, this warehouse holds objects and fragments of alien life. It's not hard to believe that the U.S. government has some secret warehouses here and there. Would they store the Ark of the Covenant in a government warehouse forever? It is not likely. *Raiders* was filmed in 1981. After all these years, wouldn't the government be using the Urim and Thummim to keep us out of wars in places such as Iraq?

PART 2

INDIANA JONES

and the
Temple of Doom

XOX XOX XOX

Giant Gongs in Club Obi-Wan

In 1935, Indiana Jones is in the Club Obi-Wan in Shanghai, dressed in a white tuxedo. A muscular Chinese man is banging a huge gong with a mallet.

Shanghai is the largest city in the People's Republic of China and the ninth largest city in the world. Shanghai is located in eastern China on the bank of the Yangtze River Delta. From 1644 to 1911, during the Ch'ing Dynasty, Shanghai was actually a small town. By the 1930s, Shanghai was divided into several areas, one controlled by the British, one by the French, and one still governed by the Chinese. When the National Party, or Guomindang, took control of China in 1912, Shanghai began to grow until it was an important center of commerce and politics. In 1921, the Communist Party was established in Shanghai. But later in the 1920s, Chiang Kai-shek unified China and drove the Communists out of Shanghai. In 1934, the Red Armies of the Communist Party went on a massive military retreat to escape the pursuit of Chiang Kai-shek's forces. This "Long March" lasted more than six thousand miles and helped es-tablish Mao Zedong as a leader of the Chinese communists. The same year, Chiang Kai-shek started a movement based on Confu-cian and Christian morals called the New Life Movement. Ideals of the New Life Movement included the notion that all women should become good housewives and mothers.

The word *gong* has its roots in Malay-Javanese, but the word is used across Asia. The type of gong used in *Temple of Doom* is called a suspended gong, which is a round piece of flat metal—usually brass or bronze—held in the air by a cord that goes through holes in the top of the circle.

The suspended gong in the movie appears to be a large version of a chau gong, also known as a tam-tam or a Chinese gong. In real life, chau gongs can be as large as eighty inches in diameter.

Dating from the early Western Han Dynasty (206 BC to AD 9), the earliest known chau gong was found in the Guangxi Zhuang Autonomous Region of China in a tomb.

The chau gong served to alert people that important processions and people were entering the area. While the slinky nightclub singer Willie Scott isn't all that distinguished, nor are her slinky dancers, in a 1935 Shanghai nightclub maybe they were important enough to warrant the gong. In the old days, the gong sounded multiple times if an incoming official was really prominent, and the number of strikes of the gong actually indicated the eminence of the official.

In Chinese orchestras that accompany operas, a pair of gongs is essential, with the smaller of the pair used for higher tones and the bigger one used for lower tones. The larger gong lets the audience know that a major character of the opera or a man is about to come on stage. It also tells people that something crucial is happening in the operatic story. The smaller gong accentuates funny things that happen in the story and indicates that a minor character or a woman is about to come on stage. This makes the giant gong amusingly inappropriate for Willie's entrance.

Anything Goes

The lovely American nightclub singer Willie Scott entertains the crowd with a Mandarin version of the classic Cole Porter tune "Anything Goes."

"Anything Goes" was from the Cole Porter musical of the same name that opened on Broadway in November 1934. It also featured the song "I Get a Kick Out of You." After 420 performances at the Alvin Theater and the 46th Street Theater, the show closed.

In Shanghai, the everyday spoken language is called Shanghainese, and about fifteen million people speak it. Although Shanghainese is widely used for verbal communication, it is rarely used in written materials. Despite its popularity, however, Shanghainese is not the language taught in schools, nor is it the most common and accepted language in China. In the early twentieth century, Mandarin was chosen as the common language platform for China. It was thought that most Chinese spoke Mandarin and that it was fairly easy to learn the language compared to other choices, which explains both why the song is sung in this language and how someone as daffy as Willie could learn the language.

The Nurhachi

When Indiana Jones enters Club Obi-Wan, he is escorted to a table where a Chinese gangster boss, Lao Che, asks him if he has found the Nurhachi. Indy responds that the night before, some of Lao's henchmen tried to steal the Nurhachi from him. Guns are drawn, and Indy grabs Willie Scott and holds a knife against her side, telling Lao to provide a particular diamond in payment. After receiving the huge diamond from Lao Che, Indy gives him the Nurhachi, which is a small container holding the ashes of Nurhachi, the First Emperor of the Manchu Dynasty.

Born in 1559 in Manchuria, the real Nurhachi died on September 30, 1626. He was indeed one of the founders of the Manchu, or Ch'ing Dynasty (1644–1911), the last imperial dynasty to rule in China, and was considered to be the First Emperor of the Manchu Dynasty.

The real Nurhachi had many different names, among them his formal title, Kundulun Khan; his reigning title, T'ien Ming; Brilliant Emperor Who Benefits All Nations, or Geren Burun Be Ujire Genggiyen; his temple name, T'ai Tsu; and Kao Huang-ti, or Chieftain of the Manchurian Tribe Chien-chou Juchen. After his death, he was called Ch'ing Taizu, or First Emperor of the Ch'ing Dynasty, when the Ch'ing Dynasty was founded.

Nurhachi's tribe, a Tungusic people, lived on the eastern border of the Chinese Empire in the Long White Mountains north of the Yalu River. Today, the largest number of Tungusic people are the ten million Tungusics known as Manchurians. The majority have assimilated into the Chinese Han population. Back in Nurhachi's time, there were five Juchen tribes in Manchuria, and his was called the Chien-chou Juchen. Nurhachi's tribe was constantly at war with the other four Juchen tribes, all of which lived farther north of the Manchurian forest and steppe region.

When Nurhachi was in his early twenties, both his father and his grandfather died in battles against the other tribes. China's Ming Dynasty had fomented the battles, hoping to make the five Juchen tribes kill each other off and hence relieving China from the need to worry about any Juchen attacks. Nurhachi's tribe was beginning to fall apart, and his own survival seemed at the mercy of fate.

Apparently, he was not the type of man to accept any fate other than one he had carved out for himself. Knowing that one of the men who wanted control of Chien-chou Juchen happened to be supported by the Chinese, Nurhachi killed the man. He took the reins and then set out to destroy all of the other Juchen tribes. Because he did not want the Chinese to support his enemies, hence giving them strong advantages over the Chien-chou Juchen tribe, Nurhachi's tactic was to first invade the part of Manchuria that was under the dominion of the Chinese. After destroying the Chinese-ruled areas, he decided that he might as well invade the entire Chinese Empire. Obviously, Nurhachi was not a man with small ambitions.

During his military adventures, Nurhachi established the Manchu state, and by 1599, the Manchurians had a writing system that later became the foundation of much literature. Then in 1601, Nurhachi established administrative and taxation organizations for the Manchurian state, thus turning a tribe into an actual empire with structures and laws. His male relatives managed the various administrative and taxation organizations, so his decisions were never challenged and he was never under threat of being overthrown by a rival family.

In addition, Nurhachi monopolized trade in ginseng, fur, and pearls, and he took charge of key mines. This gave him a huge amount of money, which is always helpful when someone is trying to take over the world—or, in Nurhachi's case, a huge country such as imperial China.

In 1616, Nurhachi declared that he was the khan, or emperor, of his new Manchurian dynasty. He actually called himself T'ien Ming, which means Heavenly Mandated, and he called his new dynasty the Chin Dynasty, sometimes called the Later Chin Dynasty because there was a Chin Dynasty in the twelfth century, also founded by the Juchen.

In 1618, Nurhachi attacked China for seven distinct reasons, which had to do with the Chinese support of the various Juchen tribes, the murders of Nurhachi's father and grandfather, and various other matters. The commander of the Chinese city of Fushun, a man named Li Yung-Fang, turned traitor to China and joined forces with the Manchurians. As a result, Nurhachi's men captured Fushun.

In 1626, the Chinese defeated Nurhachi at Ningyuan. This was his first defeat, and he died from wounds endured in the battle. After his death, the Manchu Empire grew and dominated, making Nurhachi the First Emperor of the Manchu Dynasty.

It is unlikely that Nurhachi's ashes were contained in a jade funerary urn as in *Temple of Doom*. The Yongling Mausoleum, built in 1598, is the resting place of Nurhachi's ancestors. Originally known as Xingjingling, the tomb holds Nurhachi's father, Takeshi; his grandfather Juechangan; and other ancestors. Nurhachi and his mistress Yihnaran, however, are buried in Fuling Tomb, also known as the East Tomb, in Shenyang. Huang Taiji, the father of Emperor Shunzhi, and one of Nurhachi's sons, is also buried in Fuling Tomb.

Of course, the main question about the ashes of Nurhachi remains unanswered: are those really the Nurhachi ashes in the funerary urn? Why would Indiana Jones hand over the real ashes to a gangster? How would the gangster know the real ashes from other ashes? And while we're on the subject, how would Lao know that the jade urn is the real Nurhachi urn? It wouldn't take much for Indiana Jones to substitute another urn for the real one.

Gigantic Diamonds

Willie Scott immediately forgets that her life is being threatened when she hears the word "diamond." In fact, the diamond that Indy receives as payment for retrieving the Nurhachi is enormous.

Is this gigantic diamond real? And how does Indiana Jones verify at the Club Obi-Wan that this is the diamond he really wants?

Perhaps the diamond is the Peacock's Eye from *Young Indiana Jones and the Peacock's Eye*, filmed in 1995. In the television episode,

Indy is searching for a 140-carat diamond that belonged to Alexander the Great. But this particular diamond, the Peacock's Eye, does not really exist.

Enormous diamonds, however, do exist. One of the world's largest diamonds is the Kohinoor Diamond, which once weighed 186 carats. Originating in India, where most of *Indiana Jones and the Temple of Doom* takes place, the Kohinoor was once literally the world's largest known diamond. It is thought that the diamond, an oval-cut stone, was set in the peacock throne of the Mughal emperor Shah Jehan, who built the Taj Mahal. Possibly, the Peacock's Eye of Indiana Jones is named after the Kohinoor in honor of Shah Jehan's peacock throne.

After Agra and Delhi were sacked, the diamond was taken to Persia in 1739. Then in 1747, after the assassination of the Shah Nadir, the Kohinoor landed in the hands of Afghanistan's Ahmed Shah Abdali. The diamond continued to play a role in intrigue, death, and politics. When the British took control of India, the Treaty of Lahore included a clause about the diamond, stating that it had to be surrendered to the Queen of England. Queen Victoria had the Kohinoor recut, and today it has 105 carats and is part of the British Crown Jewels. Our guess is that the magnificent diamond in *Temple of Doom* was modeled on the Kohinoor.

Another possibility for Indy's huge diamond is the Cullinan I, or Star of Africa. A much less obvious choice, this diamond was named after Sir Thomas Cullinan, who happened to own a mining company. The Cullinan is the largest known diamond in the world today. It weighs 530 carats and is in the scepter of King Edward VII in the Tower of London.

When the Cullinan was first discovered in 1905, the rough gem-quality diamond weighed 3,106 carats. It had a wonderful clarity, being a Grade D colorless gem, but it also had a black spot in its middle. The spot causes severe straining in the diamond. The Transvaal government bought the diamond and presented it to King Edward VII. At that point, it was cut into three large pieces in Amsterdam. Many other smaller diamonds were also split from the main stone. The largest piece that was cut from the original stone was the Cullinan I.

The second most massive diamond ever found is called the Excelsior, and in raw form, it weighed 995 carats. After it was cut into ten pieces, its biggest stone weighed 158 carats. The ten pieces were then cut into twenty-one diamonds that ranged from 1 to 70 carats.

The world's third largest cut diamond is the Orloff, weighing in at 194 carats. Another sizable Indian stone, the Orloff was an eye of the idol Srirangam in the temple of Brahma. Like the Kohinoor, the Orloff was acquired by the Shah Nadir. And then, unlike the Kohinoor and the Excelsior, the Orloff ended up in Russia, rather than England, where it was given to Catherine II by her ex-lover Grigori Orloff.

In case you're wondering how big these diamonds really are, 1 carat equals 200 milligrams (or 0.2 grams) of mass. So the 140-carat diamond in *Temple of Doom* weighs 28 grams, and the 105-carat Kohinoor weighs 21 grams. Now, 1 gram is equal to 0.035274 ounces, so if you're used to thinking in terms of ounces and pounds, Indiana Jones's 140-carat diamond weighs approximately 0.9876 ounces. Not particularly heavy, but *huge*: the size of a large egg.

Poisons and Their Antidotes

After giving the diamond to Indiana Jones, Lao Che holds up a vial of blue liquid, and as Willie Scott powders her nose, he tells Indy that the blue liquid is the antidote to the poison Indy just drank. Apparently, the Chinese gangsters put a fast-acting poison into Indiana Jones's martini. Lao Che will surrender the antidote only if Indy surrenders the diamond.

What kind of poison can you slip into someone's drink that will kill him within minutes? And are there antidotes to such poisons?

Keeping in mind that Indiana Jones was in China, Indy's attackers could have used one of several types of poisons in his drink. One likely candidate is *Illicium religiosum*, which was available in both China and Japan as early as 1880. In fact, in 1880 in Japan, three children died from exposure to seeds from the fast-acting poison of the killer plant. And the plant originally came from China.

With the entire plant poisonous, only one drop of *Illicium reli-giosum* is necessary to kill a person. The plant's poison has a crystal-line structure that is soluble in water or alcohol, as well as in ether or acetic acid. After ingesting the poison, someone—possibly Indiana Jones—would suffer from diminished reflexes, and he would soon have convulsions. Even a small dose causes heart paralysis.

The antidote is reportedly chloral hydrate, but when the poison is ingested in anything other than a small dose, the antidote does not work.

Another, less likely, possibility might be the blowfish, or puffer fish, which is extremely lethal. One pinch of blowfish poison can kill as many as thirty people. To kill one man, possibly Indiana Jones, only one to two milligrams is required. To put this in per-spective, this amount of blowfish poison can fit on the tip of a pin.

There is no known antidote to blowfish poison, which is as fast acting and potent as curare. Some Japanese people consider the blowfish a culinary delicacy, but when it's not prepared correctly, the gourmet fish dish kills swiftly. The blowfish is in the family Tetraodontidae, the class Osteichthyes, and the order Tetraodon-tiformes. Specifically, the genus *Fugu* lives only in Japanese waters, and the people there have eaten blowfish for hundreds of years.

Between 1868 and 1912, the sale of fugu was prohibited in cer-tain parts of Japan. Even today, health officials keep close watch on the most poisonous parts of the fish, such as the ovaries and the liver.

In the category of manmade poisons, in the United States in 1937, the taste of raspberries could mean death in the form of the elixir sulfanilamide. In only two months, the drug killed more than one hundred people in fifteen states, and it led to the passage of the 1938 United States Food, Drug, and Cosmetic Act, enabling the Food and Drug Administration (FDA) to better regulate drugs.

Sulfanilamide tablets and powder were used to treat streptococ-cal infections. Consumers in Southern states were asking for the drug in liquid form, so the chief chemist and pharmacist of the S. E. Massengill Company in Bristol, Tennessee, cooked up a liquid dose. Finding that sulfanilamide dissolves in diethylene glycol, he passed the concoction on to the production part of the company, which produced the new elixir and shipped it all over the country. Unfortunately, diethylene glycol, which is used as an antifreeze, also happens to be a deadly poison.

Nerve agents are chemically similar to organophosphate pesticides. The G-type nerve agents are colorless and tasteless liquids that dissolve in water. GB is odorless, GA tastes a tiny bit fruity, and GD smells a little like camphor. GA was first synthesized by the Germans in 1936, while GB was created in 1934.

One antidote for nerve agents is atropine sulfate, which blocks the action of the neurotransmitter acetylcholine. In modern warfare, soldiers sometimes carry autoinjectors that contain atropine.

Liquid sarin is colorless and is a cholinesterase inhibitor; it is many times more deadly than cyanide and is related to pesticides that the Germans developed in the 1930s. If absorbed through skin, a lethal 0.5-milligram dose of sarin attacks the central nervous system.

VX is another killer liquid that attacks the nervous system. It is a clear, colorless, odorless, oily liquid that dissolves in water: ethylphosphonothioic acid S-(2-(bis(1-methylethyl)amino)ethyl) O-ethyl ester. It is considered to be the most deadly nerve agent in existence and causes death within fifteen minutes. Even a tiny drop is fatal. The immediate injection of atropine may help to deter the lethal effects of VX. VX didn't appear until the early 1950s in Britain. So the choice of liquid nerve agent in Indiana Jones's *Temple of Doom* time period would probably be GB.

The Shanghai Mob

Indy and the Chinese gangsters engage in a pitched battle on the floor of the Club Obi-Wan. Indy manages to obtain the antidote to the poison he swallowed and takes it. Then he and Willie jump out a window, fall several stories, and land in Indy's car, which is driven by his young assistant, Shorty.

Was there really a Shanghai mob during the 1930s? Did they wear black suits and hats, and did they ride in American-style 1930s mafia cars? They probably didn't look like American mafia gang-sters, with the black suits and cars, but yes, they existed.

Shanghai became an international drug capital in the 1920s, so drugs to spike drinks and mobsters to control drug traffic were plentiful. In fact, the warlords were so active in Shanghai during this time that the country as a whole was becoming impoverished.

China's Nationalist movement rose to power during the period of the drug warlords. In 1924, the Nationalist Party, or Kuomin-tang (KMT), under Sun Yat-sen allied itself with the new Chinese Communist Party.

When Sun died in 1925, Chiang Kai-shek gained power of the KMT, and his control of Shanghai, an opium capital, was possible because the criminals and the wealthy people supported him. The wealthy were against any reforms that could possibly hurt their ability to make money. The criminals obviously were against re-forms, and in Shanghai they were dominated by two mobs called the Green Gang and the Red Gang.

In the 1800s, the two gangs were composed of criminals who transported grain and smuggled salt on the Grand Canal. But then, starting in approximately 1911, the Red and Green Gangs shifted their focus to Shanghai, China's largest city, and to other main cities.

Tu Yueh-sheng was a leader of the Green Gang and one of Shanghai's most powerful citizens. He was a narcotics warlord who rose to power from his roots in Shanghai's French Settlements, where criminals were allowed to do what they wanted. The French simply taxed the Chinese drug profits and ignored the criminal ac-tivities. Tu was actually under the guidance and tutelage of the French chief of detectives, Pockmarked Huang, who also served as a major Green Gang warlord himself.

After the British tried to stamp out opium traffic in Shanghai in 1918, the Green Gang stepped in and took charge of the opium trade. By the 1920s, Tu Yueh-sheng monopolized drug traffic and actually became known as the Opium King. Not only was he the Opium King, but Tu was also a heroin king because he produced an opium antidote that happened to be made out of heroin. He imported tons of heroin to produce his opium antidote.

During the 1920s, a triumvirate of mobsters commanded Shanghai's underworld, and Tu Yueh-sheng was one of them. In 1927, labor unions began a strike that included a protest against the Shanghai gangsters. By the time Chiang Kai-shek showed up in Shanghai, the people hoped that Shanghai would be free of foreign economic control and warlords. But a strong labor union was not what Chiang Kai-shek wanted. He was more interested in satisfying wealthy businessmen, both Chinese and foreign. Thus, Chiang Kai-shek delayed his foray into Shanghai and pushed the police to destroy the labor union.

The Green Gang started a reign of terror against members of the labor union and those in the Communist Party. The mob succeeded in destroying both the union and the party. At this point, with the Green Gang ruling over so many officials and so much wealth, Tu Yueh-sheng basically could do whatever he wanted, not only in Shanghai, but in a much wider range of territory.

In 1928, after the Geneva Convention banned heroin trafficking, the Green Gang in Shanghai developed its own drug refineries. By 1934, heroin use in Shanghai far exceeded opium use, and the mob was exporting heroin overseas.

By the summer of 1935, Chiang Kai-shek, wanting opium revenues for the national treasury, had given official supervision of the opium business to Tu Yueh-sheng. The government now had a monopoly on opium trafficking, with Tu Yueh-sheng, a warlord mobster, in charge of everything. Keeping in mind how important opium was to the Chinese and how addicted so many people were to drugs, a huge amount of money was being shared between the government and the mob. The government was, in essence, not only a part of the mob but its greatest ally. And indeed, Tu Yueh-sheng managed all narcotics distribution, as well as government finances and operations. He was actually the chairman of the board of directors of the Commercial Bank of China, and he was considered to be the most powerful man in the country.

In the 1935 Shanghai of *Temple of Doom*, the mob definitely would have been a major force. Given the mob's control of the police and the government, however, it is extremely unlikely that an American archaeologist, an American bimbo nightclub singer, and a young Chinese boy could escape the mob's clutches—on foot, by car, by plane, or by any means whatsoever. There would be no need for Lao Che to barter with someone as alone and weak as an American archaeologist. No need to fork over a gigantic diamond to obtain the ashes of the First Emperor of the Manchu Dynasty. No motivation whatsoever to get the ashes in the first place, much less pay for them. No need to spike Indiana Jones's drink. All Lao Che had to do was send in the police to kill Indiana Jones. Lao Che didn't even have to show up for the execution, and neither did his henchmen.

As a final note of interest, in 1937 the Japanese invaded Shanghai, and Tu Yueh-sheng left the city for Chungking. There, he was known as a wealthy benefactor of the people, a rich philanthropist.

The Route from Shanghai to India

After escaping from Lao Che, Indiana Jones and his friends Willie Scott and Shorty fly over Chungking in a plane full of live poultry. Unknown to them, the pilot releases the fuel from the plane and parachutes out of the doomed vessel. After Indy, Willie, and Shorty jump out of the plane right before it crashes, Indy saves everyone by steering an inflatable rubber raft over snow-covered mountains. He and his friends then fall off a high cliff, survive the fall, and end up floating down a river. At this point, Indy points out that they are somewhere in India.

For those of us who are not geography experts, we start to wonder how close India is to Chungking, and whether a crash landing in India is possible. We will bypass discussing the incredibly fun escapades with the inflatable rubber raft, as there is no way we can explain the survival of an archaeologist, a bimbo nightclub singer, and a young boy riding a small inflatable raft from a plane down steep mountains, over a very high cliff, and into a swamp. No bruises, no broken bones. Amazing. But it *is* an exciting, fun-filled scene!

Chungking is an extremely large municipality in the western half of China. Before 1997, it was a city in the Sichuan Province.

Shanghai is on the extreme eastern side of China, about halfway between its north and south boundaries. Following the Yangtze River through the midsection of China, you can almost trace the route from

Shanghai west over the countryside to Chungking. As a direct trajectory, as if flying in a straight line over the country in an airplane, it's approximately eight hundred miles from Shanghai to Chungking.

The distance from Chungking to the border of India is approximately four hundred miles—assuming, once again, a straight trajectory in, say, an airplane. Most likely, such a plane would end up in Burma, Bhutan, or Nepal.

India borders China, Nepal, and Bhutan to the northeast, and Burma (now called Myanmar) is east of Nepal and Bhutan. To the west of India is Pakistan. To the east are Bangladesh and Burma (Myanmar). After flying over the Himalaya Mountains in Nepal, Indy's plane could land in the low-lying areas to the north of India.

The first sea clipper introduced by Pan American was the Sikorsky S-42 in 1934, which flew from Miami to Buenos Aires and held thirty-two passengers. Its fuel tanks were barely large enough for the twelve-hundred-mile journey.

As we discussed in our chapter about *Raiders of the Lost Ark*, in late 1935, Pan Am introduced the Martin M-130, nicknamed the China Clipper. It flew across the Pacific Ocean, and in 1936, it took passengers on an eight-thousand-mile, weeklong trip to Hong Kong. Each M-130 could carry forty-six passengers.

It's also possible, though unlikely, that Indiana Jones boarded a 1934 Sikorsky S-42, although why this Pan American plane was transporting live poultry from Shanghai makes no sense. Besides, the first note of an S-42 in China is in approximately 1937. Let's use it as a model, though, since it was a cutting-edge plane of the time and was capable of making a twelve-hundred-mile journey, which is approximately the mileage required to go from Shanghai to India on a straight route.

However, there were quite a few problems on Indy's plane— chickens, exploding engines, pilots draining the fuel and parachuting out of the plane, and scrapes along the mountaintops. As pointed out, it was a twelve-hundred-mile path even in a straight line. So there is probably no way the plane could have flown in a straight path from Shanghai to Chungking, then between Nepal to the west and Bhutan to the east to crash land in the northernmost tip of India.

Let's assume, for argument's sake, that it took a total of a thousand miles to fly from Shanghai to India. The S-42 could easily make the trip. At approximately 170 miles per hour, it could fly from Shanghai to India in about six hours.

After passing Chungking, it is more likely that the plane had to travel approximately eight hundred miles (rather than four hundred miles) to reach India by flying over Nepal to the west or Bhutan to the east. To fly eight hundred miles would require four or more hours, and with the fuel drained from the plane, it is not possible that the craft could make it into India. But even if it could, for some unexplained reason, it seems unlikely that Indy, Willie, and Shorty remained on board for four or more hours after the pilot drained the fuel and bailed.

Now, perhaps the plane flew closer to the border of China before the pilot bailed. Perhaps it came almost to the border. In this case, there might have been enough fuel for the plane to have crashed in the mountains of Nepal or to the extreme north of India. If the pilot parachuted from the plane in Nepal, then it's very likely that the plane crashed shortly thereafter. Keep in mind, however, that we're talking about a kind of plane that Indy probably wasn't riding.

Several other types of plane were operating in China in 1935. For example, in March 1935, a Ford ten-passenger plane went into operation, making trips from Chungking to Kweiyarig. A Douglas fourteen-passenger plane started to fly in China on May 14, 1935, and it ran from Shanghai to Peiping, which is now Beijing. Beijing is about one-quarter to one-third of the way from Shanghai to Chungking. So while we'd like to believe that there was indeed a live poultry–hauling plane operating that was scheduled to fly and that could fly from Shanghai to Chungking and beyond, it seems far-fetched—but not beyond the pale of imagination.

Northern India and the Hindi Language

After landing right over the border in northern India, the three heroes—Indiana Jones, Willie Scott, and Shorty—encounter an old man with a shock of white hair. He points the way to a tiny village in the middle of nowhere that is surrounded by desert. The villagers descend upon the threesome, begging for money in the Hindi language. Indy, who seems to know nearly every language he encounters in foreign countries, is comfortable with Hindi and converses with the villagers and the old man.

It's possible that when the villagers provided Indy and his friends with what little food they had, he responded by saying, "Thank you," or "Dhanyavád," to them.

With as many as three hundred million Indians speaking Hindi, it is the main language of the country. It is definitely spoken in northern India, as well as throughout the rest of the country, particularly in the central regions.

In fact, Hindi is spoken in other countries, as well. Estimates place Nepalese speakers of Hindi at approximately 8 million; South African speakers at about 1 million; U.S. speakers at 300,000; and Yemeni speakers at 200,000. Hindi speakers number 20,000 in New Zealand, and speakers of the language are also found elsewhere in large numbers around the world.

Standard Hindi is derived in large part from Sanskrit. In its purest form, standard Hindi is an "official" language in India, used for public broadcasts and news, while Hindustani is used as the everyday language of the people.

Hindustani has elements of Arabic and Persian, as well as influences from English. It combines High Hindi and Urdu.

High Hindi is based on the Khariboli dialect and is written in the Devanagari script. It is considered the standard form of Hindi, which is used for the official language. In the urban areas of northern India, High Hindi is taught in schools, and many people use it on a daily basis. In the more rural areas of northern India, however, where Indy and his friends end up, the people most likely speak Hindustani, with its Arabic, Persian, and English influences.

Urdu is also based on the Khariboli dialect, but it is written in Perso-Arabic script and contains far more Arabic and Persian words than High Hindi does.

There are other forms of Hindi as well, such as Rekhta, a type of Urdu used in poetry, and Dakhini, which is more similar to Urdu than to pure Hindi.

Delhi, Pankot, and Indian Elephants

Indiana Jones wants to leave the Indian village and head to Delhi, so he asks for a guide to help him find the way. The village elder tells Indy that before going to Delhi, they must first stop at Pankot Palace, which is in a town called Pankot.

First, how far from the northern border of India is Delhi? Does it seem possible for Indiana Jones to ride an elephant all the way to Delhi?

As luck would have it, Delhi is indeed located in northern India. As we figure that Indy landed in northern India directly over the border of Nepal—giving him a fairly straight route from Chungking—we see that Delhi is yet farther north from where Indy landed. Still, it might be approximately two hundred miles between the rural Indian village and Delhi, assuming the trip is made via a fairly direct route. On an elephant, it might be possible to make the journey in about a week.

The Indian elephant, of course, is famous as a symbol of the country. It is one of three subspecies of the Asian elephant, *Elephas maximus*. Appropriately, it is known as *Elephas maximum indicus*. The Indian elephant weighs between three and five tons, with small animals weighing in at a low two tons. Even the larger Indian elephants are smallish by world elephant tonnage standards; African elephants, for example, can weigh between four and seven tons. Indian elephants could easily haul Indy, Willie, and Shorty through the deserts and the swamps and over mountains. For centuries, these animals have carried people over such rough terrains, which happen to be very difficult for horses to handle.

Delhi would be of interest to Indiana Jones for two reasons: it is a major city and hence an excellent conduit from India to other countries; and it contains many ancient artifacts, always an appealing factor to Indy.

Delhi served as the capital of many Indian empires throughout history, and today it is a very large city. It lies on the banks of the Yamuna River. In Indy's time, Delhi was probably still a major force in the Indian trade routes. Of interest to Indy the archaeologist, an area of Delhi was the capital of the Mughal Empire for a very long time, and after India broke from Britain in 1947, this area, named New Delhi, became the capital of India.

As for Pankot, it is a fictional town and does not exist in India.

The Maharajah

Indiana's guide tells him that Pankot has a new maharajah, and the palace has the power of the very dark. It is where many villagers were killed, and it is an evil place.

Would a maharajah be located in fictional 1935 Pankot, or would he live elsewhere, say, in a major city? First, it's helpful to know what a maharajah is and what the term meant in 1935.

In general, a maharajah is a great king. In Hindi, it refers to a potentate, which is a monarch, a sovereign: that is, the one and only dominant ruler.

Before 1947, with the British ruling the country, the Indian rajas and thakurs had power, but they were directly controlled by the British, who kept "residents" in the Indian palaces. A resident was a British government official with permanent residency in India. The raja was a princely ruler, and a thakur was a lower-ranking ruler, such as a feudal lord. Many other ruling titles also existed, but these were the main ones.

By 1947, when India split from British domination, the country was made up of hundreds of political states, and each state had a minor princely or feudal ruler, such as a raja or a thakur. These political states typically consisted of a town or perhaps a couple of villages. After India claimed its independence, only a few extremely wealthy and powerful states had a maharajah, or great king.

When Britain first took control of India, however, things were different. There were a lot more maharajahs before 1947, including in 1935 and extending further back. If a raja ruling even a tiny state helped the British during World War 1, for example, he might become a maharajah. Had the maharajah of Pankot Palace, Zalim Singh, supported the British, he might indeed have earned this title. And given that there is a British resident in his palace, this might make sense. One does wonder why the British resident, Captain Blumburtt, never notices the Thuggees and the evil satanic rituals, though. Of course, Blumburtt seems to be more of a guest than a resident, which would explain his innocence of the evil rituals occurring in the palace.

In real life, the Maharajah Jagatjit Singh of Punjab (it's interesting to note that the real maharajah's name was Singh, and in *Temple of Doom*, the maharajah's name was also Singh) had been a raja for seventy-one years and a maharajah for fifty-eight of those years. His palace at Kapurthala clearly equaled the fictitious Pankot Palace in opulence and luxury. In fact, what Maharajah Jagatjit Singh did for the people of Kapurthala was remarkable and far superior to anything the dreadful maharajah of Pankot Palace might have done, had he even been inclined to do so.

Maharajah Jagatjit Singh was greatly influenced by the ideas of foreigners. For one thing, while he was a raja, or minor prince, under the British rule, his state was run by men such as Sir Lepel Griffin, Sir Charles Rivaz, Sir Mackworth Young, Sir Denzil Ibbetson, Sir Frederick Freyer, and Colonel Massey. Obviously, these were all British residents in Punjab.

But perhaps more important were the maharajah's travels abroad, where he learned how to modernize his homeland. In 1893, he visited Europe for the first time, and when he returned to Kapurthala, he introduced modern sewage and water systems into his city. In 1901, he installed a telephone system so that people in different parts of his state could talk to one another. In 1904, Singh overhauled Kapurthala's judicial system, and between 1906 and 1910, he reformed the police. In 1918, he provided mandatory grade school education for his people.

These modern conveniences may have existed in Pankot and its palace, but there is no evidence of such things in *Temple of Doom*. Indeed, Pankot appears to be far removed from all modern forms of civilization. Instead of modern police, Pankot has Thuggees. Instead of an up-to-date telephone system, Pankot has a holy evil priest booming deadly commands to an audience of zombies deep within an underground cave. And as for Pankot's judicial system, it consists of arbitrary death by removal of a person's heart while he or she is still alive. This results in a zombielike state in which the victim can feel everything happening to him, followed by imprisonment within a torturous cage that plunges into a fire pit of death. Some justice system.

The real maharajahs of the 1930s weren't nearly as violent and cruel to their people as the maharajah of Pankot Palace was.

Monsoons

Before Indiana Jones leaves for Pankot, the villagers tell him that evil resides there, evil so dark and powerful that it resembles a monsoon that hits one place and sends evil all over the country.

In reality, a monsoon is an extremely rainy season with mighty winds that blow in a particular direction. In India, this season lasts for many months, with the winds driving in from the Indian Ocean and the Arabian Sea. The southwestern monsoon hits northern India, where the rural village might be located in *Temple of Doom*, from June through September.

The term *monsoon* comes from the old Arab word *mausin*, which meant "the season of winds."

Before the monsoon season, the central and northern regions of India become extremely hot. The land absorbs heat much faster than the Indian Ocean does, and the air over the land rises. During the dry season, strong winds blow from the land to the ocean, but during the wet season, the winds reverse direction.

The hot air over the land is replaced by heavier, cooler, moist air that sweeps in from the ocean. The torrential rainfall races across India from the Indian Sea, drawn toward the Himalaya Mountains. Because the mountains block the winds from entering Central Asia, the winds rise, and then the temperature falls, dumping rain on India. The layer of cold moisture over India is extremely thick, up to three miles in depth.

The Indian Ocean monsoon, or southwestern monsoon, over India is the strongest, most intense monsoon in the world.

In general, monsoons are created from three things: a difference in temperature—in particular, strong heat—between the land and the ocean; Coriolis forces caused by the Earth's rotation; and the storing and releasing of energy as water changes to vapor and then back again. We've touched on the heat difference between the Indian Ocean and northern India. The Coriolis force was first described in 1835 by Gaspard-Gustave Coriolis, a French scientist. In terms of monsoons, it basically refers to how the winds move from a straight path when in a rotating frame of reference. This force, caused by the Earth's rotation, is responsible for the direction of winds in a cyclone, for example—the winds rotate clockwise in the southern hemisphere and counterclockwise in the northern hemisphere.

So when the villagers tell Indiana Jones that the evil in Pankot Palace is so dark and powerful that it resembles a monsoon that hits one place and sends evil all over the country, what they're saying makes a lot of sense. The villagers are used to the southwestern monsoon, which hits northern India from the ocean and spreads its winds all over the country.

Sivalinga and Shiva

The villagers tell Indiana Jones that the evil people in Pankot stole the village's sacred stone, which is known as Sivalinga. This is why Shiva has brought Indiana Jones to the village, to help the people get their sacred stone back.

The Sivalinga is one of five such rocks, which together are called the Sankara Stones. They provide peace and tranquility to those who dwell near them and worship them. When the stones are in contact with one another, magical diamonds inside them make the stones glow. Each Sankara Stone has three lines cut on it, and these lines represent different levels of the Hindu universe. Unfortunately, these stones exist only as props.

The *Temple of Doom's* Sivalinga is, however, based on a type of sculpture or totem that is common in India, the Shiva Lingam (plural is Shiva Linga). This smooth carved stone is a semi-abstract depiction of Lord Shiva's organ of procreation, often paired with the corresponding female organ that is sculpted as a receptacle for Shiva's seed. According to a legend, when the renowned sage Bhrigu once went to visit Lord Shiva, he knocked on the door of the residence repeatedly but received no answer. He knocked harder, then pounded with all his strength. Finally, Lord Shiva came to the door with his wife Parvati on his arm. He had wanted to finish his lovemaking with the goddess before opening the door to the sage. Bhrigu was so furious at being treated this disrespectfully that he cursed Lord Shiva, saying that if Shiva was so fond of lovemaking, then he would forever be worshipped in the image of his organ of generation, rather than in an anthropomorphic form.

Conversely, there are more esoteric beliefs about the origin of this symbol of Lord Shiva. Yet although Shiva is visualized as an abstract primordial creative power, his followers needed to worship something tangible that represents him, hence these phallic stones were created.

Many types of "good luck" totems have been created throughout history, all over the world. In general, a totem is any physical item that people think helps or protects them. North American totem poles are one obvious example.

A very common totem around the world is that of the Earth mother or goddess. Visit any museum, and you'll see plenty of big-breasted clay figurines depicting this totem. Figures of this nature have been found dating as far back as 20,000 BC. It is believed that all human life, indeed all of Earth itself, originated from the symbolic Earth Mother and that we all depend on her.

In Hinduism, people often have shrines in their homes, where they invite gods to come and be worshipped by them. Images of gods are kept on the shrines as symbols of good fortune and luck.

In *Temple of Doom*, it seems that Shiva is a good god because he has brought Indiana Jones to the suffering village. Is Shiva a real Indian god, and if so, is he a good god or an evil one?

In real-life India, Hindu people who believe in Shiva hold the faith that there is a single Reality, as in absolute monotheism. Yet several manifestations of this single Reality exist in the form of gods. The three main gods, the Supreme Trinity of all manifestations of the single Reality, are Brahma, Vishnu, and Shiva. The term *Trimurti* (having three forms) is applied to Brahma, Vishnu, and Shiva, and the Trimurti is often depicted as a man with three heads but only one body. The three forms symbolize creation, preservation, and destruction. Everything that is good and beautiful is God, or Satyam Shivam Sundaram.

Shiva is a good god, the destroyer of all evil and the creator of regeneration of life, or reproduction. He usually appears as white with a dark-blue neck, which was caused by poison that he devoured to save the world. He has three eyes and several arms. He rides a white bull and carries a trident, or three-pronged spear. Often, he is depicted with a snake around his neck.

When he is a destroyer, he appears as a naked ascetic and his followers, which are present to help him destroy all evil, are grotesque demons and serpents. Shiva-Rudra is the destroyer of evil. When he is enabling reproduction, he is Shiva-Shankara, and as the cosmic dancer, he is known as Shiva-Nataraja.

Shiva is everything. Destroyer and creator. Youngest and oldest. The source of fertility in all life. Gentle yet fierce. For those who worship him, Shiva provides great prosperity.

Then there is Shakti, the female manifestation of God, the divine mother. She provides the energy and force for male gods such as Shiva. The female Shakti of Shiva is known by the name Parvati, and she is the daughter of Himavan, the lord of the Himalayan mountains, and Mena, a supernatural form of female menses. Without Shakti, Shiva cannot exist. Without Shiva, Shakti cannot exist. Together, they provide the bliss of One.

For the Hindu people, Parvati gives balance to Shiva, the god of destruction and reproduction. She represents marital faithfulness, fertility, devotion to one's spouse, and the power of the universe.

There are five mantras, or mystical, symbolic "thought forms," that make up Shiva's body. These are known as Eesaana, Sadyojaata, Vaamadeva, Aghora, and Tatpurusha.

The mantra Eesaana refers to the Shiva that is invisible to the human eye. The mantra Sadyojaata refers to Shiva in his most fundamental forms. The Vishnudharmottara Purana, a Hindu text from the sixth century, provides both a face and an earth element to each of the Sadyojaata fundamental forms. Fire corresponds to the mantra Aghora. Water is assigned the mantra Vaamadeva. Air is assigned to Tatpurusha, and space is assigned to Eesaana, the invisible. Because each of the five mantras has a face associated with it, it makes some sense that in *Temple of Doom*, there are five Sankara Stones.

The faces of Shiva are portrayed in stone in the Elephanta Caves near Mumba. The Elephanta Caves are on Elephanta Island, and they consist of a huge temple complex dedicated to Shiva. The

Elephanta Caves and their temples span approximately sixty thousand square feet. All of the temples are cut from natural rock; there is a main temple chamber as well as two side chambers and many smaller shrines.

The other two gods of the Supreme Trinity that make up the single Reality are Brahma and Vishnu.

Brahma is the first of the three gods in the Trimurti. He created the universe, and, in the Hindu faith, it is believed that all life evolved from Brahma. Brahma has four heads, all bearded, and four arms, and he is usually shown sitting on a lotus. He holds various things in his four hands: a *mala* (rosary), a *sruva* (religious ladle), the Vedas (divine knowledge in textual form), and a *kamandalu* (water pot). Brahma wears the hide from a black antelope, and he rides on a *hamsa*, which is a swan.

Brahma's lotus represents the single Reality, and thus Brahma is always rooted in Reality. His four faces symbolize the four Vedas, which are spiritual manuscripts. The four faces also represent the *manas* (mind), *buddhi* (intellect), *ahamkara* (ego), and *chitta* (consciousness). Together, these represent the four ways in which humans are able to think.

It is also believed that the four Indian castes are derived from Brahma's being. From his mouth came the Brahmins; from his arms came the Kshatriyas; from his legs came the Vaishyas; and from his feet came the Shudras. The caste system itself was introduced by the Aryans into India.

Originally, there were three loose castes of priests, warriors, and commoners, and the three castes had to do only with labor rather than social division. Eventually, the Aryans stopped wandering and settled into a more agricultural existence. At this time, labor became more specialized, and there were tillers and traders. The priests and the warriors were most important, however, and they received special privileges and prayers.

Another factor leading to the more formalized caste system was the Aryan exclusion of the dark-skinned Dasas from their social interactions. The Aryans refused to allow intermarriage and also restricted Aryans and Dasas from intermingling, coeducation, and having the same work functions.

The Vedas divide the people into four castes. The Brahmins are the highest caste because they were born directly from Brahma's mouth. The Brahmins request favors from the gods for all the people.

The second highest caste is the Kshatriyas, who were born from Brahma's arms. They protect the people and wield weapons.

The third caste is the Vaishyas, born from Brahma's legs. They are the farmers, and they also handle trade.

And finally, those born from Brahma's feet are the Shudras caste, and their duty in life is to serve the other three castes.

In addition to the main four castes, there was yet one more division of the people. After all, beneath those who serve the others must be a category for people deemed not worthy to serve others. These outcasts are called the Chandalas, and they are not even allowed to enter the village, much less live in it.

The final manifestation of the Trimurti is Vishnu, the preserver of the universe. Vishnu is goodness and mercy. He is dark and, like Brahma, he has four arms. His hands also hold symbolic items: a conch shell, a club, a ciscus, and a lotus. He wears yellow robes and is often shown resting on a coiled, thousandheaded snake named Seshnaga, which is floating on the cosmic ocean.

The Three Lines of the Universe

To help the villagers, Indy, Willie, and Shorty decide to head for Pankot Palace.

The sacred stones in the film each have three lines cut into them, and these three lines represent the three levels of the universe. In Judeo-Christian theology, these might be heaven, the Earth, and the underworld. In *Temple of Doom*, the three lines might represent (1) the transcendental level, in which the only Reality is Brahma; (2) the pragmatic level, or the material world; and (3) the apparent level, in which the material world is an illusion. Or they may represent the three manifestations of the single Reality, the Trimurti of Shiva, Brahma, and Vishnu. After all, the Trimurti is a man with three heads on one body, and the three manifestations represent creation, preservation, and destruction.

Giant Vampire Bats

While traveling on elephants from the rural village to Pankot Palace, Indiana Jones, Willie Scott, and Shorty stop for the night to camp. Willie is attacked by giant vampire bats, which are also seen in great flocks later when they arrive at Pankot Palace.

Most likely, the movie included giant bats because they are feared by most people as symbols of doom and gloom, as portents of evil things to come.

There are more than a thousand types of bats in the world. A bat, in general, is a mammal whose forelimbs are wings that enable it to fly.

The word *bat* is derived from the Old Norse word *ledhrblaka*, which means "leather flapper." Over time, this word was shortened from *ledhrblaka* to *bakka*, and from there, it became the even shorter *bat*. As for the word *vampire*, it comes from the Magyar word *vampir*, which actually means "witch."

Most bats are not vampiric. In fact, 70 percent of all bats feed on insects. Almost all of the other bats eat fruits.

Only three species of bats feed on blood. These are the common vampire bat, the hairy-legged vampire bat, and the white-winged vampire bat.

Vampire bats have short muzzles, small ears, and short tail membranes. Their teeth are built for cutting into flesh, and their digestive systems are specialized to handle blood. Rather than sucking blood, the vampire bat laps it after secreting draculin, an anticoagulant, into the victim. Draculin (we are not making up the name!) is a glycoprotein consisting of 411 amino acids.

True to the Dracula myth, vampire bats hunt only at night. It does not make much sense at all that Indiana Jones would see a flock of, say, two hundred gigantic vampire bats flying over Pankot Palace in the daylight. But it does make sense, assuming these bats are in India, that they might try to suck Willie Scott's blood at night in the campsite.

The only type of bat that likes the blood of mammals such as Willie Scott is the common vampire bat. The other two blood lappers, the hairy-legged vampire bat and the white-winged vampire bat, feed on the blood of birds.

The usual prey for a common vampire bat is a sleeping mammal. So the vampire bats in *Temple of Doom* would be less likely to attack Willie Scott than they would any other mammal sleeping elsewhere. Remember, Willie is fully awake and messing around with her clothes after washing up.

Typically, the common vampire bat lands on the ground and does *not* fly into the hair and the face of its victim. Rather, it sneaks up on the sleeping animal. It either walks or runs to the victim.

To drink the blood, first the common vampire bat cuts the fur from the animal's skin using its teeth. If the prey is a human, this probably doesn't happen, although, in the case of an excessively hairy man, perhaps the bat still has to shear some of the "fur" from the skin. The razor-sharp incisor teeth cut into the skin, and at that point, the vampire bat injects draculin-drenched saliva into the wound. The tongue laps the blood while the draculin keeps the blood from clotting. Thus, the bat gets to feed longer than it would without the draculin.

During one twenty-minute feeding, a common vampire bat can consume one ounce of blood. Immediately, its stomach lining absorbs the blood plasma, which has no nutritional value. The kidneys pass the plasma to the bladder, and within a few minutes after lapping blood for dinner, the bat excretes the plasma. At that point, the vampire bat goes home to its roost, relaxes, and digests blood for the rest of the night.

As for bats flying during the day in flocks of one to two hundred, this seems unlikely. Vampire bats prefer completely dark locations—caves are a good example. While there may be several hundred bats in one colony, it would be unusual to see them flying in a flock like birds migrating for the summer.

As for size, these are not gigantic bats. The common vampire bat might have a wingspan of eight inches. Its body is small, nowhere near the fist-size of the huge vampire bats in the film. It has no tail, and the total length of its head and body together might measure three inches.

Now for the big question: do common vampire bats live in India? And the big answer is, no.

The common vampire bat lives in an area ranging from Mexico to Chile. The other two types of vampire bats, both of which feast on insects, are found in Texas, eastern Mexico, and southern Brazil.

In Africa and Asia, there are "false" vampire bats that eat birds, frogs, and small mammals. The *Megaderma lyra* "false" vampire bat lives in the area from eastern Pakistan to Sri Lanka, in southeastern China, and in the northern Malay Peninsula. Pakistan is to the west of India. Sri Lanka is near India but not on the continent. Rather, Sri Lanka is separated from India by the Palk Bay and the Gulf of Mannar. If you look at a map, you will see that Pakistan borders the west of India, while Sri Lanka is to the southeast of India. Hence, the "false" vampire bat does indeed live in India itself. Of course, southeastern China is also close to India, given that China borders the north of India. The Malay Peninsula is south of Cambodia and Thailand, which are across the Indian Ocean from India.

How likely is it that an Asian "false" vampire bat attacks Willie Scott? It's true that these bats live where *Temple of Doom* takes place. Like real vampire bats, however, the false variety is nocturnal and sleeps during the day, so again, seeing a flock of hundreds of them in the daylight is unlikely. And again, unless the bat that attacked Willie Scott was insane, it would not want to feed on her blood. While the *Megaderma lyra* is a carnivore, it eats insects, spiders, birds, fish, and rodents. If it enters someone's house, it will eat insects off the walls rather than lap up the person's blood for supper.

Baboons in India

After being attacked by a giant vampire bat, Willie Scott is frightened again, this time by a baboon.

Each attack on Willie by another animal is meant to raise the audience's level of general merriment. It's hard to take her predicament seriously when we see Willie Scott startled by one giant creature after another. We can't quite believe that all of these beasts are truly dangerous and out to get her. As Willie flutters around the camp, unable to get help from her companions, the audience is increasingly amused.

In the order Primate and the family Cercopithecidae, baboons are ground-dwelling monkeys. There are two classifications within Cercopithecidae. The *Chaeropithecus* consists of several types of baboon: the largest baboon, which is the black-gray chacma in eastern and southern Africa; the yellow baboon in central and southern Africa; the brown doguera baboon in east central Africa; and the reddish-brown western baboon in west central Africa. There is only one member of the genus *Comopithecus*, the other classification of baboon within Cercopithecidae. It is the sacred Anubis baboon, otherwise known as the hamadryas. It lives in Egypt, Saudi Arabia, Somalia, Ethiopia, and Sudan.

If there was any type of baboon in India in 1935, as portrayed in *Temple of Doom*, it might have been the sacred Anubis. This baboon lives farther north than any other baboon and was sacred to the ancient Egyptians, who thought that the *Papio hamadryas* (aka hamadryas or sacred Anubis) was the attendant of their god Thoth, one of their most important gods.

The average weight of an Anubis baboon is about forty pounds. It has coarse body hair and naked pads on its buttocks. The adult male has side whiskers and a big mane. The baboons have long canine teeth, powerful jaws, and strong arms and legs. The females are half the size of males. And the two sexes have completely different appearances: the females are brown and don't have manes, and the males are silvery white with manes.

On a dark night, particularly after being attacked by a vampire bat (or possibly a "false" vampire bat), Willie Scott might have been terrified to see a large baboon by the campfire light.

Baboons are social animals, and they live together in groups that can have as many as fifty members. When they travel together, the most dominant males stay inside the center of the group. The mothers and the infants also stay in the center, for protection. Along the perimeter of the circle of baboons are the young adults, who are agile and strong and can warn the group of approaching danger, if needed.

In the wild, baboons rarely fight among themselves. The most dominant males generally have the largest canine teeth, and they hardly ever battle for ultimate dominance. The rest of the baboons follow suit. If the dominant males don't fight one another, then the other baboons don't squabble, either.

It's true that a baboon might attack an unarmed human. Although baboons tend to eat during the day and sleep at night, sometimes they roam around at night, attacking young farm animals such as lambs. But a baboon's typical diet does not consist of lamb and human flesh. Mainly, it eats plants, small animals, and scorpions.

Because baboons roam together and sleep at night, however, it's not too likely that Willie Scott would run into a lone baboon at her campsite.

Giant Owls of India

Willie Scott then encounters a giant owl, which elicits even more screams from her.

Again, Willie's fear is shown on film to keep the audience amused by her predicament. Most likely, a giant owl will not eat her, bit for bit, bone for bone, tooth for tooth. But at this point, *anything* will scare poor Willie.

In many cultures, owls represent ghostly apparitions, the underworld of death. Shamans have tried to use the supernatural powers of owls.

In India, people think owls are messengers of bad fortune, servants of the dead, and portents of evil things to come. Several types of owls are found in sacred areas. The *Strix leptogrammica*, or brown wood owl; the *Bubo zeylonensis*, or brown fish owl; and the *Bubo nipalensis*, or forest eagle owl, all live in dense forests that tend to be considered sacred groves. Hindu legends claim that when the forest eagle owl calls from cemeteries and sacred groves at night, it means that the spirit of a dead person has left the physical world. The Hindu goddess of wealth, Laxmi, rides on an owl, and certain people today believe that it's a sign of impending great wealth if an owl comes into someone's house.

We're not sure why the American bimbo lounge singer Willie Scott is afraid of Indian owls. She does not seem to be practicing the Hindu faith, nor does she seem aware of any Indian customs, the language, and so on.

Big Reptiles of India

After becoming alarmed by so many jungle creatures, Willie Scott is terrified again, this time by a big reptile with a brown, flicking tongue.

India is full of reptiles, including cobras and kraits. There are also plenty of crocodiles and other species. The gharial, *Gavialis gangeticus*, which resembles the crocodile, lives in rivers. Lizards are everywhere, as are turtles. So certainly, Willie could have been startled by a big lizard in the Indian forest.

In the movie, the reptile resembles a baby Komodo dragon (*Varanus komodoensis*). In 1912, some pearl fishermen anchored their boat on a remote island called Komodo in a string of islands called the Lesser Sundas. The fishermen claimed to have seen huge, prehistoric animals on Komodo. Up until this time, people thought that large lizards had become extinct thousands of years ago.

The Buitenzorg Zoological Museum in Java sent an expedition to Komodo and published its findings. The expedition had found the huge prehistoric creatures. The report went unnoticed.

Then in 1926, the American Museum of Natural History sent an expedition led by W. Douglas Burden to Komodo. Explorers in this expedition reported that they had discovered gigantic lizards that were more than ten feet long. The male lizard was so fierce and strong that it could kill a water buffalo that weighed several times more than the lizard's body weight. Two live lizards and twelve dead ones were brought back to the United States from Komodo Island.

Although the Komodo dragon is a dangerous lizard from the human standpoint, it lives only on Komodo Island and a few neighboring islands. So, while the lizard in *Temple of Doom* resembled a young Komodo dragon, it could not possibly be this type of lizard.

It might seem that Willie Scott encounters an iguana; however, these large reptiles are not common to India. Which leaves us with the possibility of the agamid lizard, which looks a lot like the iguana, and the agamid has the advantage of being native to India. The agamid has spiny scales, a large head, and long limbs, all of which are seen on Willie Scott's lizard.

The Blanford's rock agama is a type of agamid lizard that lives throughout India. In fact, its name is derived directly from William Thomas Blanford, who belonged to the Geological Survey of India.

Indeed, there are many species of agamid lizard in India. The Indian spiny-tailed lizard, *Uromastyx*, is another lizard in the group Agamidae. The male spiny-tailed lizard may be as large as 1½ feet, nearly the size of the lizard seen by Willie Scott.

Snake Charmers

While arguing with Indiana Jones, Willie Scott mistakes a giant snake for an elephant's trunk, causing her to scream once more.

The snake is a creature that India has no lack of. In addition to the cobra, other venomous and deadly snakes in India include the saw-scaled viper, Russell's viper, the king cobra, and the common krait. In schools, children are taught how to identify killer snakes and what to do if someone is attacked by one. Of the four hundred species of snakes in India, approximately one-fifth are poisonous.

The Indian cobra, the *Naja naja*, is usually associated with snake charmers. When threatened or angry, it raises the front part of its body and spreads the hood on its head. It tends to be anywhere from three to six feet long.

The Indian cobra is intertwined with life in India, and the Hindu god Shiva (see our earlier section about Shiva) is shown with a cobra around his neck. And, as we mentioned earlier, Vishnu is often portrayed as reclining on Seshnaga, the giant snake with one thousand cobra heads.

India also has no lack of snake charmers, with tens of thousands plying their trade. Typically, the snake charmer blows a pungi, or horn, and the snake rises up and seems to dance. In actuality, the snake does not hear the horn at all. It senses vibrations in the ground from the tapping foot of the snake charmer.

Snake charmers tend to use deadly cobras, although sometimes they use pythons, and some have been known to stage fights between a snake and a mongoose to entertain their audience. Most likely, the snake is not so charmed by the event. The mongoose usually wins.

In 2004, snake charmers in India threatened to release cobras into the state assembly at Bhubaneshwar. Apparently, the government claimed that the snake charmers tortured the cobras, and authorities seized hundreds of cobras from snake charmers and released them into the wild. The protestors, mainly from the snake charmers' village of Padmakesharpur, demanded that the government stop interfering with their right to earn a living. The government, however, refused to give in, and snake charmers were forced to find new ways to earn a living.

The Mutiny of 1857

In the banquet hall of Pankot Palace, Maharajah Zalim Singh, a young boy, enters, and a general discussion ensues about the "Mutiny of 1857."

In *Temple of Doom*, Pankot Palace apparently played an important role in the mutiny of 1857. Was there really an Indian mutiny in 1857 against the British? Most definitely. While the British may have actually called it the Mutiny of 1857, many Indians instead refer to it as the Indian Rebellion of 1857 or the First War of Indian Independence.

Whatever its official name, the Mutiny of 1857 began in early 1857 and ended in the middle of 1858. The Indians were fighting against the British East India Company's control over them. This control began in December 1600 under Queen Elizabeth I of England. She gave the British East India Company a royal charter to do trade in the East. In 1608, British ships docked in Surat, India, and by 1612, due to British support of battles against the Portuguese, the Mughal emperor Jahangir was favoring the British in trade and other matters. In 1615, King James I sent an ambassador to Jahangir's palace, and a contract was established that allowed the British East India Company to build trading posts in India, where trade was conducted for tea, silk, indigo, and other items.

By the middle of the 1600s, the British East India Company had factories in major cities all over India, including Calcutta and Bombay.

Then in 1670, King Charles II allowed the company basically to become its own nation. The British East India Company had its own army, made its own money, and governed the legal and administrative matters in all of its territories. The company ended up ruling over the Bengal presidency, the Bombay presidency, and the Madras presidency. These vast regions of Indian territory were now owned and controlled by the British.

During British rule, from 1769 to 1773, ten million Bengalis died from famine, and then later in the late 1800s, still under Britain's reign, forty million Indians died from famine.

In 1773, the British government created a new position, that of governor-general of India, a clear and major step toward claiming that India was part of Britain. The first governor-general of India was Warren Hastings, who served from 1773 to 1785. In 1787, Hastings was impeached for corruption, but, as is often the way in politics, he was acquitted of the crime in 1795.

Charles Cornwallis, the First Marquess of Cornwallis and the Second Earl of Cornwallis, was the second man to serve as governor-general of India. He forged agreements with Indian tax collectors employed by the Mughals to collect money from the peasants.

For half a century, the British actively quashed any Indian rivals to their power. By 1800, the British East India Company had expanded its territorial holdings as far as southern India. By 1850, the company fought numerous wars to further extend its territory. By 1856, the Indians were furious about the British takeovers.

The Indian soldiers, called sepoys, were grumbling about caste discrimination. Although the British had encouraged caste distinctions and the sepoys had risen in status, this high status was now being threatened. In addition, the soldiers received very low pay, and as new areas of India were annexed, the soldiers were no longer paid extra wages for foreign missions to those areas. Perhaps most significant, the British had become evangelical Christians and were trying to convert the Indian sepoys to Christianity.

Then in 1857, a new type of rifle was supplied to the Indian soldiers. This rifle only exacerbated the Indians' increasing unrest. They felt that the British were browbeating them to convert to Christianity. The cartridges that came with the rifle were supposedly lubricated with pork lard, which is considered unclean as food by the Muslims, and with beef fat, which is considered sacred by the Hindus. To load their new rifles, the soldiers had to bite open the paper cartridges containing the gun powder.

In addition to the Indian soldiers' growing fury, the Indian aristocracy was also rising up in revolt. After all, the British had taken their power from them. And if that weren't insufferable enough, the British were using a so-called Doctrine of Lapse, which stated that any Indian ruler without a male heir relinquished the land under his control upon his death. The land went to the British East India Company. Thus, many people in the noble families, some of whom came from long lines of Indian land holders, found that they were reduced to having a very low status in their own country. Emperors were forced from their own palaces, and when the British publicly auctioned the jewels of the royal family of Nagpur, Indians saw the auction as a mighty slap in the face, indicating a lack of respect of their people by the British.

Clearly, the fact that the British East India Company was making a fortune from India's silk, gold, and jewels was enough to infuriate many people. Jewels were being auctioned off, and silk and gold were being used to satisfy tax payments. Food prices had risen for the Indians. And all of the wealth the British accumulated from the Indians was being funneled into building Britain and financing British conquests in other countries.

On May 9, 1857, the Indian sepoys in the British Indian Army began to revolt against the British. The first thing they did was refuse to load their rifles with the cartridges that were greased in pork lard. Eight-five Indian soldiers were publicly humiliated, stripped of their uniforms, and sent off to do ten years of hard labor. Because their crime was refusing to do something that directly contradicted traditional Indian religious beliefs, other Indian soldiers stood behind them.

The growing unrest resulted in the full-scale Mutiny of 1857. In May, the Indian sepoys in the British Indian Army revolted outside of Delhi. They marched into Delhi, offered to fight for the Mughal emperor, and initiated a rebellion that lasted for more than a year. While many Indian kingdoms and soldiers fought against the British East India Company, some elected to fight alongside the British.

There weren't many combatants: two thousand fully equipped British troops fought twenty-three hundred sepoys who had no artillery, but that imbalance of gun power lulled the British into complacency. Because the British military authorities felt that they could easily quash a rebellion of this type, they neglected to react quickly enough, and the Indians won the first major victory of the war.

At the end of the long and bloody war, everyone of importance in the British East India Company was dead, and control of the company went to the British government. Unfortunately for the Indians, however, the British government stepped in and took control of India, and the native people, in village after village, city after city, were massacred by the British. The British remained in control for another ninety years.

Why Did It Have to Be Snake Surprise?

A luscious banquet is served, and much to Willie's and Shorty's horror, the food features native delicacies.

First to be served is "snake surprise," which consists of huge snakes wrapped around silver poles on trays. The servants slit open the snake skins, and lo and behold, extremely long, living wormlike things slither out; one of the guests eats the long wormlike things alive. Next, the servants bring in platters of large black beetles. Willie's soup contains eyeballs. Dessert is chilled monkey brains, with the guests lifting the tops off whole monkey heads and dipping their spoons directly into the craniums.

The West has long presented Eastern customs as bizarre and fantastic, but the banquet scene in *Temple of Doom* is so over the top that it can only be a send-up of this tendency. Of course, Indians do not eat anything remotely like what was served.

There are reports floating around of certain ethnic groups eating monkey brains throughout the ages, but under close inspection many of these tales turn out to be hearsay, in which the person telling the story always names *another* ethnic group, not his, as having this dietary practice. Some of the countries named in these stories are Indonesia, Vietnam, Cambodia, Taiwan, China, Hong Kong, Japan, Thailand, India, Africa, and the Middle East. The most common report is about ethnic Chinese Indonesians. The practice originally seems to have Chinese origins, going back to two historical documents. One, from the eighteenth-century Ch'ing Dynasty, is a menu called "Man Han Quan Xi" ("The Manchu-Han Complete Banquet") that supposedly features the most exotic foods in the Chinese Empire. The other report describes a general who traveled around China, possibly in the sixteenth century, and a feast in which he ate monkey brains. This document, "ManTuoLuo Xuan XianHua" ("Casual Chat on Mantuolou's Veranda"), was written by Zhang HaiOu in the nineteenth century.

Yet according to some accounts, it appears that this dietary practice does still occur in modern times. A reporter for the *Los Angeles Times* wrote a 2003 article describing Indonesian restaurants that serve live monkey brains in the same manner as in *Temple of Doom*:

> Some establishments serve macaque at a special table with a hole in the center. The monkey is tied up and the top of its skull cut open with one slice of a sharp knife. The animal, still alive, is placed under the table so its head protrudes like a bowl. Arrack, a powerful native alcohol, is sometimes poured into the skull and mixed with the brain.[1]

Note that according to this report, the practice occurs in Indonesia, not in India. From most accounts, modern-day India would never condone eating monkey brains. As for what Indians *do* eat, let's take a trip back in time.

The earliest Indians ate wheat, lentils, and rice. The wheat was baked into flatbreads called chapatis, which are still made today. These early Indians were known as Harrapans, and they came to India from Africa in approximately 40,000 BC. By 4000 BC, they began farming, and by 2400 BC, they had learned how to irrigate their fields. Sometimes, the early Indians ate animals such as chickens, cows, sheep, and pigs.

By 300 BC, many Hindus believed that animal sacrifices prevented people from being reincarnated. As a result, these sacrifices became less common. Many Hindus became vegetarians. Those who were not vegetarians ate less meat than they had in the past.

Then, in approximately AD 320, India was ruled by the Gupta Empire. King Chandragupta II united all of northern India and parts of southern India under his reign. He and subsequent Gupta kings were Hindus, and around AD 650, the Hindus started to worship the mother goddess in her various manifestations, such as Parvati (see our earlier discussion about the many gods of India). The mother goddess held cows to be sacred animals. As a result, Hindus stopped eating beef.

Still later, in approximately AD 1100, Islamic peoples invaded India. Arab troops conquered northern India and what is now called Pakistan. These Arabs had already triumphed over Persia. Many Hindus converted to Islam during this period, and given that the Koran forbids the consumption of pork, many Indian people stopped eating pork as well as beef. Even more people turned to vegetarianism than before, although some Indians still ate chicken and lamb.

The Indian diet became a combination of wheat flatbreads, chickpeas, rice, vegetables, spicy vegetarian sauces, yogurts, and meats such as chicken and lamb. It remains this way today.

The Brahmins, for instance, are strict vegetarians who eat fish in certain areas.

In northern India, because the weather alternates between scorching hot and freezing, the food is heavier than in the southern part of the country, so the people tend to eat more meat. In the north, they also eat a large variety of breads, including chapatis, naan, puris, rotis, and phulkas.

In the dry regions of India, there aren't many fresh vegetables, so the people substitute preserves and dals. Again, the foods are heavily spiced.

Had the Pankot Palace banquet been strictly accurate, we might have seen the people eating from metal plates called thali, with bowls arranged on them. The bowls would contain various spicy vegetarian foods. One bowl might hold rice, another might have bread.

An ordinary meal of a farmer might consist of yogurt, bread, and shredded vegetables. The vegetables are mixed with spices and rolled in the bread, with the resulting stuffed bread called parathas. Some people eat meat dishes, such as the famous tandoori foods, which are meats that have been marinated with spice pastes and then roasted in a tandoor, or primitive clay pot.

In Kashmir, rice predominates. Rather than consume meat, people tend to eat lotus roots and morel mushrooms. They drink a spicy green tea called kahava, which is poured from a samovar, or a large metal kettle. Fish, lamb, and chicken are sometimes eaten here, as well. A banquet in Kashmir called a waazwaan consists of thirty-six courses, none of which include monkey brains or worms.

All over India, the diet is primarily the same, with regional variations based largely on spices and the amount of meat and fish consumed. In some regions, bamboo shoots are popular. In other locales, coconut milk is used in cooking meals. There are bananas, kebabs, and sweet desserts. There are prawns and duck dishes. But there are no monkey brains and worms.

In real life, even the worshippers of Kali (in the movie, the Thuggees) were Hindus who ate vegetables. They never ate snakes or monkeys. Shiva, as you might recall from an earlier section in this part, is typically shown with snakes wrapped around his neck. Hanuman, the monkey god, is revered in the Hindu religion.

Of course, there are many cultures in the world for whom snake meat is a delicacy. And Indians have been known to eat giant fried locusts and consider them delicious. In fact, for many cultures, eating insects is a common practice. Ten thousand years ago, our ancestors ate bugs simply to survive. The ancient Romans and Greeks enjoyed beetle larvae, and today many people in Asia, Africa, and

Latin America still consume insects as part of their daily diets. The Chinese are known to eat bee larvae, the Japanese to eat fly larvae, and elsewhere, people consume meals of dragonflies, bee larvae, moth larvae, butterfly larvae, cockroaches, tarantulas, cicadas, ants, caterpillars, grasshoppers, termites, agave worms, and grubs. Some Africans eat giant black beetles. In the Philippines, people eat June beetles and water beetles, as well as ants, crickets, grasshoppers, katydids, dragonfly larvae, and locusts. In Asia, giant waterbugs are very popular.

Eyeball soup is relished in some places, although maybe not in India. In China, there are indeed fish eyeball soups, and in South America, soup often consists of boiled sheep heads, including the eyeballs, in broth. Supposedly, there are places in Istanbul that serve sheep eyeballs in thick cream sauce.

Giant Insects

In a hidden passageway in Pankot Palace, Indiana Jones and Shorty, and later Willie Scott, step on piles of crunching things that turn out to be thousands of maggots, enormous spiders, centipedes, beetles, and scorpions.

Were these giant insects created for the purposes of the movie, or do such insects really exist in India?

With more than a million species of insects on Earth, there are certainly some gigantic creepy crawlers digging their way through forests and deserts. The Goliath is the biggest beetle in the world, and it also happens to be one of the largest insects. It can grow from 2 to over 4 inches long. It is found only in the tropics of Africa, however, mainly near the equator. The Acteon beetle is also huge, measuring several inches long, several inches wide, and about 1½ inches thick, and it lives in South America. The giant weta of New Zealand is extremely heavy and, when pregnant, can weigh 1½ ounces. In captivity, it sometimes grows to 2½ ounces or more, but, as noted, these beetles are found in New Zealand, not in India. There are giant African stick insects that grow up to 14 inches long.

As for winged insects, nothing in the world tops the impressive size of the Queen Alexandria birdwing butterfly, with a wingspan of up to 12 inches. This butterfly lives in the remote jungles of New Guinea, however, not in India. The Hercules moth, also known as the Atlas, can have a wingspan that measures eleven inches, but it lives in southeast Asia.

In India, there are walking sticks, beetles, and all sorts of longish bugs. The Indian walking stick isn't enormous, however. In fact, most Indian bugs aren't as big as kittens. According to one report, most insects in India are approximately one quarter of an inch long, with the bigger ones being the size of a bean.[2] The same report points out that while it's unusual to see an oversized insect, the locusts tend to be the exception in India. And there is also the oleander hawk moth with its wingspan of up to about 4 inches. Also, the tree-boring beetle is rather large, as is the black-headed cricket. These are fairly substantial insects, but even a 4-inch bug isn't nearly as horrific as the mouse-sized insects in the film.

Thuggees and Kali

When Indiana Jones ventures down the secret passageways in Pankot Palace, he eventually comes to a ledge that arches over an immense fire pit, a platform with a huge Kali idol on it, and a platform on the other side of the fire pit with hundreds of chanting men who are bowing and kneeling. The room is lit with torches. The Kali statue has eyes and a mouth of fire. The Thuggee high priest, Mola Ram, comes out on the stage wearing a headpiece with horns.

Apparently, the Thuggee cult was decimated a century before this film takes place. The Thuggees worshipped the evil goddess Kali, we are told, using human sacrifices. Kali was one of the evil manifestations of the mother goddess and a consort to Shiva.

In actual Hinduism, there is indeed an evil goddess Kali, who is one of the many manifestations of the Hindu mother goddess and who serves as a consort to Shiva. And central to the entire movie of *Indiana Jones and the Temple of Doom,* Thuggees did indeed roam in India, robbing, torturing, and killing people.

Let's talk about the Thuggees first. The word itself in Hindi means "thief," and the Indian *Thuggee* during British rule was the precursor to the English word "thug." That should give you some idea as to the nature of the real Thuggees. The word also means "conceal" in Sanskrit. The Thuggees were a mob of sorts; its members lived throughout India in secret, networked groups. They robbed, tortured, and murdered people. It has long been believed that the Thuggees, after operating for centuries, finally petered out in the 1900s, so by 1935 they were probably not congregating. As pointed out in *Temple of Doom*, however, the Thuggees hadn't been seen or heard from in a century, and then, lo and behold, they popped up again in the caverns beneath Pankot Palace. But did they really drink blood, turn into zombies, and worship Kali?

As portrayed in the film, the real Thuggees were known to the wealthier citizens of India. They weren't merely highway robbers. Rather, they often befriended their wealthy victims, who just happened to be traveling. After gaining the confidence and trust of a wealthy victim, the Thuggee strangled, robbed, and buried him.

And yes, the strangulation conformed to rituals related to Kali. A yellow scarf was used, however—not a large, flattened cage. Yet it must be recognized that the Thuggee murderers did indeed kill according to ancient religious rituals related to Kali, and these rituals included blessings over the axes that were used to dig the victims' graves. After the murder and the robbery took place, the Thuggees held ceremonies in honor of Kali and gave the idol some of the robbed goods.

Kali, the goddess of destruction and death, was indeed behind all of this mayhem. According to Thuggee lore, a long time ago a demon named Rukt Bij-dana ate people as soon as they were born. Kali tried to kill Rukt Bij-dana, but every time the goddess was able to spill a single drop of the demon's blood, a new demon was created from the drop. Soon, there were far more demons than Kali could handle. She grew exhausted from battling all of the demons. So she paused and brushed the sweat from her body. Two human men rose from Kali's sweat.

Kali gave each of the two men a scarf, or rhumal. The scarves were to be used to strangle demons and then passed down to the next male generation in the families. If any man who didn't belong to the Kali-worshipping family used the strangulation scarves, the followers of Kali were commanded to kill him. Every murder became a sacrifice to Kali, under her orders.

The murders themselves were somewhat in accordance with what we see in the movie. We say "somewhat" because most of the movie is not particularly faithful to Thuggee practices. What *is* accurate is that the Thuggees believed that by murdering people on behalf of Kali, they were able to keep her quiet for another millennium. By strangling people, the Thuggees were performing an honorable, noble, religious duty. They did not feel immoral or evil.

Also, the Thuggees used their own language, called Ramasi, for chanting and even for talking. In this way, they could recognize each other in faraway regions of India even if two particular Thuggees had never met before. And oddly enough, because the murders had heavy religious overtones, the government accepted the Thuggees as a legitimate entity, and Thug was a bona fide profession in India.

Although the numbers of Thuggee murders are truly unknown, it is thought that a man named Behram, a member of the Thuggee cult, strangled 931 people using the Kali yellow scarf. Estimates vary from 50,000 up to 1 or 2 million in total strangulations by the Thuggees.

The Thuggees did not perform their murders in huge caverns lit by dozens of torches, though. Instead, they lured their victims to dark places that the murderers knew well. While the victims were being strangled, other Thuggees played music or made a lot of noise. Quite often, the corpses were thrown into wells—this is not quite the same as live people being lowered into fiery pits, as in *Temple of Doom*.

The high priest in the movie might have been the leader of one of these Thuggee gangs. If so, his real title would have been *jamaadaar*.

Unlike the film's portrayal of the Thuggees, the real killers prohibited the shedding of blood. They thought that Kali wanted them to strangle people and to kill without spilling blood.

In addition, the Thuggees did not use young boys to do their dirty work. To become a Thuggee as an adult, a boy did not undergo hypnosis or drink blood from a skull. In fact, nearly all Thuggees had to be born into the cult. Boys as young as ten years old would watch their fathers, uncles, and other close adult friends attack and strangle victims. This is how they learned the proper procedures. By the age of eighteen, they started to make their own human sacrifices to Kali.

In 1822, an officer in the Bengal Army named William Sleeman got the job of eradicating the Thuggee cult from India. It was Governor-General Lord Bentinck who appointed Sleeman for the task. He was a good candidate for the job: he was fluent in four Indian dialects and the first British official to prove that the mass murders throughout India were being performed by the Thuggees.

Sleeman marked all of the known Thuggee murder sites on a map, along with the dates of the murders. Eventually, he was able to predict where the Thuggees might attack their next victim. Sleeman sent his agents to the predicted location of the next attack, and when the agents noticed some known Thuggees in the area, Sleeman sent policemen there disguised as wealthy merchants. The Thuggees would attack the policemen, only to find themselves outmaneuvered. Within a decade or so, Sleeman, his agents, and his police were able to round up close to four thousand Thuggees. Approximately fifty Thuggees supplied the names of other cult members, the locations of murders, and other vital information. All of the other cult members were either hanged or imprisoned for the rest of their lives. In keeping with their Kali traditions and beliefs, those who were hanged typically asked whether they could put the nooses around their own necks.

Ripping a Heart out of a Live Body

After Mola Ram, a high priest, arrives, some zombielike slaves drag a young man onto the stage. They lock the man in a cage. The priest chants, reaches toward a bunch of skulls on the wall, then thrusts his hand into the man's chest, ripping out his heart. Oddly enough, the heart is still beating. The hole in the man's chest automatically closes, and somehow he is still alive.

The zombies crank some wheels, and they lower the caged prisoner into the fire pit. The man is burned alive, and his heart simultaneously bursts into flames in the hands of the Thuggee high priest.

Is it possible to rip someone's heart from his chest, leaving him still alive? The obvious answer is no. A person, even a high priest of the Thuggee cult, cannot stick his hand into a man's chest, rip out the man's heart, and leave the man still alive—with his chest miraculously sewn up as if his heart had never been ripped out at all.

That said, there are ways to keep "heartless" men alive for short periods of time. Consider heart transplant surgery, whereby a doctor removes a diseased or damaged heart and replaces it with a healthy one. After cornea, kidney, and liver transplants, the heart transplant is the next most common form of transplant operation in the United States.

If a person is brain dead but still has a functioning heart, that healthy heart can be put into a solution that keeps it alive for a while. Then another patient with a bad heart is put into a deep sleep and the doctor cuts into his chest, circulates the patient's blood through a machine to keep it healthy, and removes the bad heart. The doctor stitches the healthy heart in place in the patient's body, disconnects the machine, and monitors the patient as blood flows through the new heart into his body.

It is by no means a simple procedure. It cannot occur within thirty seconds or one minute. The surgeon cannot rip the bad heart from the patient's body. And if things are not carefully monitored in the operating room, the patient can die.

So it's true that the heart can survive for a while without a body and that the body can survive for a while without a heart. It's not true, however, that the heart can be ripped from the body, with the chest skin instantly repairing itself as if it has not been wounded, and it is also not true that the body would remain alive as long as shown in *Temple of Doom*.

Drinking Blood

When Indiana Jones, Willie, and Shorty are captured by the Thugees, Indy and Shorty are put in a cage with a young Indian boy, who tells Indy that the Thuggees make people drink the blood of Kali, and then the people fall into the "black sleep" and become Thuggees. The people, the boy

explains, are alive but in a nightmare that never ends. In short, when
people drink the blood of Kali, they become zombies.

We've already established that the real Thuggee cult was against
the shedding of blood. The actual Thuggees did not drink blood
from skulls.

In *Temple of Doom*, however, the screenwriters were obviously
familiar with Tibetan history and culture. Their use of "skull cups
of blood" can be traced to ancient Tibetan Buddhist texts and silk
paintings and, earlier than that, to either the Tibetan Bön religion
or ancient India. Tibetan iconography does depict certain wrathful
deities holding skull cups of blood, and the chants or liturgies of
some Tibetan Buddhist ceremonies make reference to the symbolic
drinking of blood from skull cups. Metal facsimiles of skull cups
(from tiny to life size) can be purchased from Tibetan art dealers,
and some of the older ones are made with actual human skulls, set
in ornate metal carvings.

This does not mean, however, that Buddhist practitioners actually drank blood. In ceremonies, the skull cups were filled with
liquor or another liquid to symbolize blood. It may seem strange to
Westerners that a religion using this type of symbolism could produce such a peaceful, compassionate leader as the Dalai Lama. Remember, however, that Christianity has its own strange symbolism,
of drinking the blood and eating the body of Christ in the form of
wine and a Communion wafer.

Traditionally, those who drink blood are vampires and cannibals. We'll discuss each form of blood drinker, starting with the
vampire.

Vampire lore has been around since ancient times. People in
cultures all over the world have believed in these blood-sucking
creatures. The ancient Chaldeans in Mesopotamia believed in vampires, as did the ancient Assyrians, who wrote about vampires on
clay and stone tablets.[3]

In China, vampires are often portrayed as red-eyed monsters
with green hair. In ancient India, vampire legends were evident
from the paintings on cave walls of blood-drinking creatures. In
certain writings in 1500 BC, the destroyer Rakshasas is depicted as a

vampire, and paintings from 3000 BC show the Lord of Death drinking blood from a human skull. The Indian Baital vampire is a mythological monster that hangs upside down from trees, much like a bat. The Baital hasn't any blood of its own.

From the Far East, vampire lore spread from China, Tibet, India, and the Mediterranean to the coast of the Black Sea, and from there, to Greece and the Carpathian mountains: Hungary and Transylvania.

Some of the richest vampire lore comes from Russia, Bulgaria, Serbia, and Poland: the Slavic people. The word *vampir* is related to the Russian word *peets*: "to drink." When the Slavs migrated from north of the Black Sea, they converted to Christianity. During the ninth and tenth centuries, the Eastern Orthodox Church and the Western Roman Catholic Church battled for control of Christianity. In 1054, the two churches formally separated from each other, and the Russians, the Serbians, and the Bulgarians went with the Eastern Orthodox Church, while the Croatians, the Polish people, and the Czechs went with the Roman Catholic Church. The Eastern Orthodox faction decided that the living dead were vampires.

But in *Temple of Doom*, there are no vampires. Nobody sucks blood from anybody else. It's possible, however, that cannibalism is at the root of the practice of drinking blood from human skulls.

Simply put, cannibalism refers to the eating of human flesh by other humans. While we don't see the fictional Thuggees eating flesh, we might assume that they practice cannibalism based on their practice of drinking human blood, not to mention the nasty bones and skulls mounted in templelike structures in front of the palace.

The practice of ritual, or sacred, cannibalism with religious overtones has occurred all around the world. Cannibals have lived in New Zealand, Africa, Australia, the Amazon, North America, Central America, Lithuania, and elsewhere. Cannibalism remains one of the most disgusting, feared, and ultimate taboos in society.

When Christopher Columbus traveled to the West Indies, he reportedly encountered the Carib West Indies tribe eating other humans. Calling the tribe by the name "Canibs," the explorers inadvertently created the term *cannibalism*. In Spanish, the word *canibales* came to mean "thirsty and cruel." In English, *canibales* was translated into the word *cannibalism*.

In Mexico, the Aztecs ate thousands of humans every year. They ate the members of other tribes as well as members of their own tribes. They supposedly felt that eating human flesh created a balance between the earthly world and the rest of the creation, or what lies within the expanse of the cosmos. By eating other people, the Aztecs appeased the gods, who would destroy all humanity if people were not sacrificed to them. To the Aztecs, cannibalism also gave their holy men a method of communicating with the gods.

As late as the 1960s, tribes in New Guinea were practicing cannibalism for similar reasons. Scientists found that the cannibals had contracted something akin to mad cow disease and that they passed the disease to their children at birth. The disease, known as kuru, was a direct result of eating human flesh and, in particular, of eating human brains.

Zombies

When the Thuggee high priest forces Indy to drink Kali blood from a skull, Indy appears to become a zombie. He goes mad, twisting and tortured on the ground, as if the evil blood has affected him. His hands shake like crazy and then fall still. He stares vacantly.

The Thuggees bring Willie onto the stage to be sacrificed. The high priest's hand comes down to rip out her heart but moves instead to Indy. The priest wants Indy to rip out the heart instead. At this moment, Shorty burns Indy with a torch, and Indy instantly snaps out of his zombie trance.

In our world, the notion of zombies originated with Haitian voodoo culture. In fact, the word *zombie* comes from the Haitian word *zombi*, which means "spirit of the dead." As the story goes, voodoo priests called *bokors* studied enough black magic to figure out how to resurrect the dead. But first they put their victims into a near-death state using a powder called *coup padre*.

The primary ingredient of *coup padre* is deadly tetrodotoxin from the porcupine fish, the fou-fou. The tetrodotoxin disrupts communication in the brain and is five hundred times more deadly than cyanide. A tiny drop of tetrodotoxin can kill a man.

This weird poison, *coup padre*, was made by first burying a bouga toad (called a *Bufo marinus*) and a sea snake in a jar. After the toad and the snake died from the rage of being confined in the jar, the *bokor* extracted their venom. The toad's glands held bufogenin and bufotoxin, each being from fifty to a hundred times more deadly than digitalis. The bufogenin and the bufotoxin increased the victim's heartbeat. In addition, the toad's glands held bufotenine, a powerful hallucinogenic drug.

To these drugs, the *bokor* added millipedes and tarantulas to tcha-tcha seeds (which caused pulmonary edema), nontoxic consigne seeds, pomme cajou (cashew) leaves, and bresillet tree leaves. Both of these types of leaves are related to poison ivy. Having ground everything into a powder, the *bokor* buried the concoction for two days, after which he added ground tremblador and desmember plants, two plants from the stinging nettle family, which injected formic-acid-like chemicals beneath the victim's skin; and dieffenbachia, with its glasslike needles, which made the victim's throat swell, causing great difficulty in breathing and talking. The *bokor* then added the sharp needles of the bwa pine.

He next added poisonous animals to the deadly powder. Two species of tarantulas were ground up and mixed with the skins of white tree frogs. Another bouga toad went into the mixture, followed by four types of puffer fish, the fou-fou carrying the *coup padre*. The final ingredient was human flesh from a cadaver

If a family or a community hated someone sufficiently, they called upon the *bokor* to turn that person into a zombie. After ingesting the *coup padre*, the despised villager or family member immediately became numb. His lips and tongue went numb first, followed by his fingers, arms, toes, and legs; then his entire body went numb. He became weak, and nauseous and suffered from vomiting, diarrhea, stomach pain, and headaches. The victim's pulse rapidly picked up, he had trouble walking and talking, and finally paralysis set in: his breathing became shallow, his heart nearly ceased to beat, and his body temperature plummeted. The victim's body was now blue, his eyes were glassy. In essence, the victim was in a coma.

Although the poor victim was still alive, he was buried as if already dead. Because he wasn't really dead, the victim often heard his own funeral and was horrified to suffer through his own burial.

Later, the *bokor* dug up the body and brought the person back to life. Physically, the person appeared as he did before ingesting the *coup padre*, but mentally, his mind was gone and his soul was dead. Being traumatized, the victim believed that he had been reanimated, brought back to life. As a mindless drone, this new zombie remained under the *bokor*'s power and did the *bokor*'s bidding. The *bokor* then gave his new zombie a hallucinogenic mixture of Datura stramonium, cane sugar, and sweet potato. There is absolutely no antidote for tetrodotoxin, so once a zombie, always a zombie.

In the Haitian culture, the zombie has no human consciousness or free will—just as Indiana Jones and all the children who drink the ritualistic Kali blood supposedly have no human consciousness and free will. The relationship between the Thuggees and Kali worship and Haitian zombies is never explored in the film. In reality, there is no true correlation between the two. Yes, zombies have existed in our real world, although they were "created" as we just described. The "real" zombies did not turn into soulless drones by drinking blood from skulls.

Voodoo Dolls

While Indiana Jones battles the Thugees after freeing all of the enslaved children, the Maharajah plunges a knife into a voodoo doll, making Indy writhe in pain. When Shorty pulls the knife from the voodoo doll, Indy snaps out of his painful writhing and beats up the remaining Thuggees.

Just as zombies are more a Haitian notion than an Indian one, so is voodoo. It is a faith that traditionally existed in West Africa and has migrated to other places, such as New Orleans, where it has a prominent cult following.

Along with the zombies depicted in *Temple of Doom*, voodoo is connected to satanic practices. Neither makes a whole lot of sense in Hindu culture.

Voodoo dolls are not particularly common in India. They are more of a European folk magic phenomenon, which somehow migrated into popular notions of the voodoo faith. The voodoo doll represents the spirit of a person, and you can use the doll to make the person do things. Supposedly, if you stick a pin into a doll, you curse the individual associated with that doll.

Although the voodoo doll is rooted in European folk magic, it remains a symbol of evil and unrest even today. In 2006, Chinese officials banned the sale of voodoo dolls, which were becoming increasingly popular with teens. There were voodoo dolls with names such as "The Magic Shadow Killer," which destroyed people's spirits should the owners stick pins in the dolls. And there were also loving voodoo dolls such as "Little Angel," which brought the owners love and good fortune. Teens had to buy several dolls to perform various services against many different people. Each doll could be used to perform only one service with one person in mind. The dolls were not made in New Orleans or Africa; rather, they were manufactured in Thailand. In banning the sale of these dolls, the Chinese government stated that the voodoo dolls promoted feudalism and superstition.

Crocodiles

Our heroes escape the Thuggee lair and make their way to a wooden bridge suspended high above a river. Crocodiles swim beneath Indy, Willie, and Short Round.

We mentioned earlier that crocodiles do indeed exist all over India. For example, the marsh crocodile (*Crocodylus palustris*) is found in coastal saltwater lagoons, as well as in freshwater lakes, marshes, and ponds.

The mugger crocodile lives in India and was common there as far back as the early 1900s. It is also found in Pakistan, Sri Lanka, Bangladesh, Nepal, and Iran. It prefers freshwater lakes, rivers, and marshes: shallow waters are best.

The muggers have very broad snouts, and like other crocodiles, they nest in holes, with each female laying approximately thirty eggs per year. Adult muggers eat monkeys and other small mammals, as well as reptiles, fish, birds, and amphibians.

The Indian mugger crocodile's numbers have greatly diminished, mainly due to killings by people who wanted their skins. In 1972, the mugger, along with the saltwater crocodile and the gharial, landed on India's list of endangered species. Being on the protected species list means that the crocodiles are once again flourishing in India.

Estimates are that African Nile crocodiles kill approximately one thousand people every year. The Australian saltwater crocodile is as deadly as the African Nile crocodile, but because the former tends to live far away from humans, it causes fewer deaths. The Indian muggers are also extremely dangerous, but they have dwindled dramatically in number. In fact, only the Nile crocodiles cause large numbers of human deaths.

Oddly, India's greatest predator, the tiger, isn't featured in the Indiana Jones movie at all. Worldwide, tigers kill approximately seven hundred people per year. It is estimated that only five thousand or so tigers remain in the wilds of the entire world, with half of them in India.

PART 3

INDIANA JONES

and the
Last Crusade

XOX XOX XOX

Utah, 1912, and the Boy Scouts

In 1912, a teenage Indiana Jones is a Boy Scout living with his father, an archaeologist. On a field trip with his troop to the desert, Indy stumbles across a group of grave robbers who have just found a long-lost artifact, the Cross of Coronado.

Working under the assumption that everything in the world is connected, that every detail of the Indiana Jones chronicles exists for a reason, we immediately ask, Why Utah? Why 1912? And why the Boy Scouts? While this adventure helps to shape Indy's character and opinions, there's no question that every occurrence in a man's life affects his personality. What makes these specific incidents, exciting as they might be, so memorable to Indiana Jones that a quarter of a century later he's still tracking down the Cross of Coronado? In the vast chronicle of Indiana Jones's life, what makes this day so special? Let's examine the facts and see what conclusions we can deduce. After all, the key to knowing a man is understanding his past.

The geographical area of the United States known as Utah was home to tribes of hunter-gatherers who flourished in the Great Basin of the Salt Lake during a period from 10,000 BC to AD 400. From these tribes in northern Utah emerged Native Americans who became known as the Fremont Culture. These Indians remained hunter-gatherers but also raised crops of maize, beans, and squash and lived in masonry homes. In addition, the Fremont tribes made pots and clay figures for use in their religious rites.[1]

During the same period that the Fremont Culture appeared in northern Utah, the Anasazi Indians began to settle in southern Utah. The Anasazi built rectangular buildings into the sides of cliffs and thus became known as cliff dwellers. They also made clay figures, gray-black pottery, and twined baskets. A peaceful people, they depended on crops of maize, beans, and squash. The word *Anasazi* means "the ancient ones" in Navaho.

The Anasazi flourished until approximately AD 1200, when changes in the climate brought about crop failures and attacks from hostile Indian tribes. A commonly held theory has the Anasazi migrating south and integrating with the Pueblo Indians of Arizona. In 1912, however, these theories were years in the future. Instead, a shroud of mystery enveloped the cliff-dwelling Indian tribes of Utah. Many historians of the period theorized that the Anasazi disappeared almost overnight, the victims of some unimaginable catastrophe. Modern techniques for determining climate change and tribal movement were not available, and the mystery of the lost Indian tribes made good reading.

Several early-twentieth-century anthropologists, including Frank Hamilton Cushing and Alfred V. Kidder, theorized that the Anasazi didn't vanish but merely became part of other tribes. Still, it's unlikely that a young Indiana Jones, living in a frontier outpost fifty miles from nowhere, was familiar with their work. Instead, most likely he was a reader of the popular fiction of the time. And one of the most notable themes of novels aimed at teenage boys was that of the lost race, as typified by the work of H. Rider Haggard.

Haggard, a British author who had spent a number of years in Africa, wrote *King Solomon's Mines* on a dare from a cousin, with Haggard claiming that he could write an adventure novel as entertaining as *Treasure Island*. The book told of the discovery of the biblical King Solomon's diamond mines in the heart of Africa by a party of hunters led by scout Allan Quartermain. The hardcover was published in England in September 1885 and was an immediate best seller. It was the first adventure novel written about Africa, and it was also the first novel dealing with the discovery of a "lost race," the name given to a secret or hidden tribe living in an out-of-the-way location never before explored by Europeans.

The huge success of *King Solomon's Mines* made Haggard rich and launched him on a long career writing lost-race novels that took place all across the globe. At the end of the nineteenth century and the beginning of the twentieth century, many regions of the world were still unknown to American or European explorers.

Within months of the publication of Haggard's novel, half a dozen more lost-race novels appeared. In one year, there were twenty such books. In ten years, a hundred. By the time Indiana Jones learned to read, there were dozens of cheaply produced books dealing with lost races of Indians living in the great forests of the northwest, the Grand Canyon, and all over Canada. The mysterious cliff dwellers made for good copy, and Nick Carter, the preeminent boys' detective character of the time, stumbled upon lost races and beautiful princesses two or three times a year. It didn't matter that a logical explanation existed for the disappearance of the Anasazi. The first rule of writing escapist fiction was to ignore any facts that might hinder a good story.

For the record, although by now it's been well documented that climate changes and marauding Indian tribes caused the Anasazi to abandon their cliff dwellings, those accepted facts haven't stopped modern-day novelists from speculating on their "mysterious" disappearance. Louis L'Amour, the best-selling Western author of all time, scored with *The Haunted Mesa*, a novel that speculated that the Anasazi used interdimensional travel to escape from their enemies. Douglas Preston theorized about the Anasazi disappearance in his book *Thunderhead*, and Kathleen O'Neal and W. Michael Gear wrote an entire series about the Anasazi disappearance that they called *The First Americans*. It's easy to see why a young Indiana Jones would become fascinated by archaeology when living near such incredibly interesting ruins.

The mystery of the disappearing Indians must have caught and held the vivid imagination of young Indiana Jones, just as he must have been fascinated by the unusual history of his home state, Utah. For, despite Utah having been accepted as a member of the Union by the U.S. Congress in 1894, that wasn't the first time the settlers had petitioned the government to join the Union. There was the curious episode of the giant state known as Deseret.

The earliest foreigner to set foot in Utah was most likely American fur trapper Jim Bridger, sometime in the nineteenth century. While Francisco Vázquez de Coronado, the Spanish explorer, is mentioned in passing in *The Last Crusade*, there's no real evidence that he actually ever made it to Utah during his unsuccessful search for the legendary city of gold, Cibola. Fur trappers, including Jim Bridger and Peter Ogden in the 1820s, hunted and mapped in the territory that later became the state. It wasn't until 1847 that a large number of settlers arrived in the Salt Lake Valley. These people were members of the Church of Latter-Day Saints, known as the Mormons. At the time, the land was still controlled by Mexico. Soon afterward, however, the Mexican-American War took place and the territory became part of the United States under terms of the Treaty of Guadalupe Hidalgo in March 1848.

The Mormons proved to be aggressive settlers, and within a few years of their arrival at the Great Salt Lake they had established a number of other communities in the surrounding territories. New settlements included Bountiful, Ogden, Tooele, and Provo. In 1851, Fillmore was established with the aim of making it the capital of the Utah territory. Meanwhile, Mormon soldiers and missionaries set up outposts in Las Vegas, Nevada; San Bernardino, California; Fort Lemhi, Idaho; and other points throughout the far West.

In 1849, the Mormons petitioned the U.S. Congress for statehood. The name of the state was Deseret, as chosen by Brigham Young, the leader of the Mormons, based on a reference from the *Book of Mormon*. The state of Deseret included all of Utah, along with large sections of Idaho, Nevada, Wyoming, Arizona, Oregon, New Mexico, and California. Congress was hesitant to allow the establishment of such a huge state, especially one controlled and mostly populated by members of the Church of Latter-Day Saints. Many representatives worried that too much territory would be controlled by too few people. Others were concerned about the difficulty the army would have in patrolling the state. No one even asked about Mormon religious practices, since, at the time of the original proposal to create Deseret, little was known about the Mormon practice of polygamy.

In 1850, the Utah Territory was established by Congress, with Fillmore, Utah, as the capital of the territory. In 1856, the capital was changed to Salt Lake City. Salt Lake City was the final link of the First Transcontinental Telegraph line, which was finished in October 1861. Brigham Young, the leader of the Mormons, was one of the first people to send a telegraph cross-country. On May

10, 1869, the First Transcontinental Railroad was completed at Promontory Summit, Utah. The train brought numerous immigrants west seeking their fortune, greatly expanding the population of the territory.

In 1854, the Mormon practice of polygamy was made public through a series of newspaper articles and was greeted with overwhelming negative publicity throughout the United States. The Mormons were considered traitors and savages by most of the population of the country. America was a strict Christian nation, and multiple spouses were just not allowed. It wasn't until the Church of Latter-Day Saints banned the practice of polygamy with the Manifesto of 1890 that Congress once again was willing to consider Utah's application as a state. After so many men spent forty years trying, the territory officially became a state on January 4, 1896.

While Utah was not the edge of civilization at the beginning of the twentieth century, neither was it the center of modern life. Despite Indy being involved in a wild automobile chase in the Young Indiana Jones chronicles, chances are that any actual pursuit in real-life Utah would have taken place on horseback, not via motorcar. In 1909, there were 370,000 people living in Utah, but they owned a grand total of only 873 cars and trucks. The high cost of automobiles was just too much for most citizens. It wasn't until Henry Ford perfected the assembly line in 1913 that Model Ts became affordable and Americans took to the road.

Still, Utah was a fascinating place for a boy to grow up, and it seemed quite likely that the presence of the cliff-dweller ruins and the unusual history of the state were two strong factors that shaped a young Indiana Jones's interest in archaeology. But there was a third influence on Indy, and without question, this occurrence prompted many young men to seek a career in the field of archaeology. It was an event of historic significance that took place in 1912. This event turned out to be a hoax.

We've already determined that Utah, with its unique history and Anasazi legacy, was a special place for a boy interested in archaeology to live. But what of the time? What made 1912 so special? In 1912, New Mexico was admitted as the forty-seventh state of the United States. Roald Amundsen announced his arrival at the

South Pole on December 14, 1911, a story that carried over into 1912. The Girl Scouts of America were founded (the Boy Scouts of America had already been established two years earlier) in 1912. The unsinkable ship the *Titanic* sank.

The 1912 Olympics took place in Stockholm, Sweden. American Jim Thorpe won the pentathlon and the newly created decathlon. Years afterward, Thorpe had his medals stripped away because he had played semiprofessional football before the Olympics, a violation of the amateur athlete code. In 1982, thirty years after Thorpe's death, his medals were restored to him. Unfortunately, no one was able to restore Thorpe's life, which ended in abject poverty. In 1950, the Associated Press ranked Thorpe the greatest athlete of the first half of the twentieth century.

The discovery that swept the sinking of the *Titanic*, the heroics of Jim Thorpe, and the finding of the South Pole off the newspapers' front pages took place in a small town in England. It was there, in 1912, that the most famous archaeological hoax of all time began. The event was the discovery of pieces of a human skull and jawbone in a gravel pit in Piltdown, England. The Piltdown hoax was important for two major reasons: it focused public attention on human evolution, and it was the longest-running hoax from its initial announcement to its exposure as a fake.

The "Piltdown Man" tale began simply enough. At the center of this strange story was a man named Charles Dawson, who in 1908 received a bit of human skull that had been found in a gravel pit in the village of Piltdown near Uckfield, Sussex. Intrigued, Dawson visited the gravel pit many times over the course of the next four years, finding additional pieces of the skull. He finally sought expert advice from the head of the British Museum's geology department, Arthur Smith Woodward. After examining Dawson's skull pieces, Woodward proclaimed that Dawson had discovered the remains of a completely new subspecies of early man. Together, Woodward and Dawson returned to Piltdown and found additional skull chips in the gravel pit.

Experts at that time felt that the men had indeed discovered the fossilized remains of a new subspecies of man. They called the new human *Eoanthropus dawsoni* ("Dawson's dawn-man" in Latin). But it was easier for most people to think of the new human as Piltdown Man.

Using the fragments of the skull, a reconstruction of the head of the Piltdown Man was developed. Studying the completed skull, Arthur Smith Woodward theorized that the Piltdown Man was the missing link in evolution between ape and man. He based this notion on the skull's humanlike cranium combined with the apelike jaw. Woodward never seemed to suspect that the two very different bone fragments indicated that they actually came from a human and an ape.

Not all scientists were as convinced as Woodward of the authenticity of the Piltdown Man. For example, Franz Weidenreich, a human anatomy expert, stated that the bone fragments were actually a modern man's skull and an orangutan's jawbone. But it was exciting to believe that a missing link between man and ape had been found, and so nobody paid any attention to Weidenreich's conclusions. The world wanted to believe that a link between man and ape existed, so it did.[2]

It wasn't until 1953 that the Piltdown Man was finally exposed as a fraud. Joseph Weiner, a professor of anthropology at Oxford University, assembled a solid body of evidence that proved conclusively that the Piltdown Man was a cleverly constructed fake. By now, the study of evolution had advanced sufficiently that scientists realized that the existence of a missing link made no sense. The Piltdown Man didn't fit into the evolutionary chain, based on numerous other human fossils found elsewhere in the world. Thus, Weiner's conclusion that the Piltdown Man was a fake, unlike Weidenreich's, was greeted with a sigh of relief by anthropologists and archaeologists everywhere.

While many people blamed Charles Dawson, who found the bones, as the person behind the hoax, no one was able to say for sure, since Dawson died in 1916, long before the forgery was discovered. At least sixteen other people were blamed for the forgery, even Sir Arthur Conan Doyle, the author of the famous novel *The Lost World*, about finding a plateau in South America where dinosaurs still existed. The worldwide attention generated by the discovery of the Piltdown Man must have been noticed by a young and impressionable Indiana Jones. Perhaps, along with the mystery of the Anasazi and the fascinating history of his home state, the Piltdown Man discovery was enough to convince a young Indiana Jones that he wanted to be an archaeologist. Like many such notions, it remains a mystery.

It's no mystery why Indiana Jones hated the Nazis. The party of Adolf Hitler, the leader of 1930s Germany, stood for bigotry, intolerance, and racial hatred, all beliefs that were anathema to patriotic Americans. Yet strangely enough, Indy was once associated with the hated symbol of the Nazi party, the swastika. Not by choice but by circumstance.

The Boy Scouts were the creation of Robert Baden-Powell, a lieutenant general in the British Army. Baden-Powell was stationed as a military officer in India and Africa during the 1880s and 1890s. Since his childhood, he had been fond of woodcraft and military scouting, and in the wilderness he taught his soldiers how to survive in the harsh outdoors. In doing so, he observed that such training helped his men to develop independent thinking instead of merely following their officers' orders.

In the Great Boer War in South Africa, Baden-Powell was given the task of holding the town of Mafeking with a small garrison against a much larger Boer force. With his troops outnumbered five to one, Baden-Powell came up with several innovative ideas to hold the fort. The most successful of these was the establishment of the Mafeking Cadet Corps, a group consisting of teenagers from the town who carried important messages behind the lines in a time before the invention of field telephones. Every member of the Cadet Corps received a badge that consisted of a combined compass point and a spearhead.

In the United Kingdom the public followed the battle of Mafeking through the newspapers, and when the two-year siege was broken, Baden-Powell became a national hero. That status boosted sales of a book he had written about his experiences teaching soldiers the rudiments of military scouting. When Baden-Powell returned to Britain, he was surprised to learn that his book was being used by teachers and youth groups for boys. A number of people suggested to him that he rewrite the book for young men.

Incorporating ideas that were suggested to him by naturalist Ernest Thompson Seton, Baden-Powell wrote a new book, *Scouting for Boys*. Published in England in 1908, the volume soon became known as the *Boy Scout Handbook*. It went through numerous printings all over the world and became the fourth best-selling book of all time. The Boy Scout movement, based on Baden-Powell's book, spread swiftly throughout the British Empire. Troops formed all over England, and within months, scouting organizations were started in Gibraltar and Malta. Canada became the first overseas dominion with a sanctioned Boy Scout program, followed by Australia, New Zealand, and South Africa. By 1910, there were authorized troops in nearly all of Europe, South America, and the United States. The first Boy Scout rally, held in 1910 at the Crystal Palace in London, attracted ten thousand boys.

In the United States, scouting used images drawn from the U.S. frontier experience. A sanctioned Boy Scout troop was formed in Utah in 1911. No doubt this was the group to which Indiana Jones belonged on that fateful day in 1912. In an odd twist of fate, it's quite likely that Indy first encountered the swastika through scouting. The symbol was used on the Thanks Badge, created in 1911. The swastika had been a symbol for luck in India long before being adopted by the Nazis, and Baden-Powell would have come across it during his years serving in that country. In 1922, the swastika was incorporated into the design for the Medal of Merit. The symbol was dropped by the Boy Scouts in 1934 because of its use by the Nazi Party. Still, it's quite likely that Indy, as a good scout, was awarded a Thanks Badge in 1912, an unusual foreshadowing of the menace he would face later in life.

Grave Robbers, Spanish Conquistadors, and the Cross of Coronado

Believing that such important relics as the Cross of Coronado belong in museums, not in private collections, young Indiana Jones steals the cross from the grave robbers.

When a teenage Indiana Jones encounters a group of grave robbers in the Utah desert, it's unlikely he realized that a quarter-century later, he'd be engaged in pretty much the same activities. Not that Indy would ever admit to such a link, at least not to anyone other than his friend Marcus Brody. But, despite protests to the contrary, the only difference between grave robbing and early acts of archaeology is who ends up with the loot. In the case of grave robbers, it's a wealthy patron, while for archaeologists it's a wealthy institution.

Basically, grave robbing is the act of locating a tomb, a crypt, or an otherwise sacred burial plot to steal the artifacts inside. In some cases, such activities might include stealing the body or the skeletal remains in the tomb or the clothes, the jewelry, or the personal possessions buried with the corpse. Archaeology, as discussed in our chapter on *Raiders of the Lost Ark*, is the study of human culture through the examination of relics from past societies. In its earlier years, archaeology focused mostly on the study of ancient civilizations, particularly those of Greece, Rome, and Egypt. Archaeology was considered a part of history. In modern times, archaeologists have concentrated on the study of human society over the ages. Archaeology is now considered one of the four branches of anthropology.

The looting of ancient burial sites by people searching for buried treasure is a problem that has existed for thousands of years. The ancient Egyptians were among the first people to believe that after a man died, he passed on to another existence, and that he was able to take as much of his wealth and property as he was able to have buried with him.

The earliest surviving records and tombs from ancient Egypt indicate that the very wealthy had themselves interred with jewelry, gold, precious stones, weapons, table utensils, and clothing. Many of the most powerful priests and leaders of the kingdom were buried in jewel-encrusted sarcophagi, with solid-gold masks laid across their faces. When a pharaoh died, no expense was spared to make his stay in the afterlife as comfortable as possible.

Inscriptions dating back to the Old Kingdom of Egypt have been found on tomb walls, warning that grave robbers would be judged harshly by the gods in the afterlife. Extremely painful punishment also awaited any thief caught looting a tomb in this earthly life. Still, the gold and jewelry proved to be too much of a temptation for many thieves. Moreover, it's been theorized that many tombs were looted by the priests who handled the final burial arrangements, as well as by the tomb builders. A number of ancient tombs were swept of gold before they were even sealed.

Tombs were designed to discourage entry by grave robbers. Subterranean burial chambers had entranceways that were blocked by huge stone slabs that slid into place when the tomb was closed. The entrance to the tomb was then barricaded by stones and rubble.

Pyramids, which served as tombs for powerful kings, included a number of blind passageways that led nowhere. These tombs also featured numerous trapdoors that opened onto deep, deadly shafts. A series of slabs descending from the roof of the major passageways helped to deter theft by falling on thieves and crushing them. Yet somehow, nearly every known tomb in Egypt was looted. Certain tombs were looted more than once. After each theft, the Egyptian priests restocked the tomb with artifacts and resealed the chamber, only to have the thieves strike again and again.

With the decline of Egypt's civilization and the rise of burial rites that didn't require any sort of treasure to be buried with kings or nobles, grave robbing settled down to a long period of relative inactivity. Artifacts still surfaced in the marketplaces of London and Paris, but widespread looting of ancient tombs seemed a thing of the past—until the arrival of Giovanni Battista Belzoni.

Born in the Italian city of Padua in November 1778, Giovanni never received a formal education, although he did learn how to read. When he was sixteen, Belzoni left home and set out for Rome. There, according to his memoirs, he studied hydraulics. For the rest of his life, Belzoni liked to refer to himself as a hydraulics engineer. Unfortunately, there wasn't a big demand for such engineers, so Belzoni turned to acting. A huge man, well over six feet tall, Belzoni was famous for his immense strength. He appeared in many plays, usually taking the role of a giant or a cannibal king.[3]

Belzoni spent the next decade touring the British Isles and performing stunts involving feats of strength. In 1803, he married a woman named Sarah, who accompanied him on his later trips all over the world. In 1813, Belzoni and Sarah toured Spain and Portugal. In 1814, they traveled to Sicily, and from there to Constantinople.

It was in Constantinople that Belzoni heard that the ruler of Egypt, the Pasha Muhammad Ali, wanted to modernize his country and was recruiting engineers despite their lack of experience. In May 1815, Belzoni and Sarah set sail for Egypt. In Cairo, Belzoni promised to build a giant water wheel for the Pasha. He and Sarah and their servant, James Curtin, were given a house on the grounds of the Pasha's palace, and Belzoni was awarded a small salary. While in Cairo, the Belzonis acted like typical tourists.[4]

After a year, Belzoni's water wheel was finally finished, but an accident occurred during the demonstration of the wheel, with James Curtin breaking his leg. The Pasha refused to pay for the wheel because he figured it was too dangerous. Belzoni found himself stuck in Egypt without money or a job.

He soon learned that the British consul in Cairo was in the market for Egyptian relics. At the time, the Egyptians had no interest in the statues and the sarcophagi from thousands of years ago. They broke many of the artifacts and used the crumbled bits as building material. Nor did Pasha Muhuammad Ali care about the relics—not as long as the people digging for antiquities paid for a digging permit. And, needless to say, the museums of Europe were quite happy to pay the minimal cost for a permit if it allowed them to cart away relics of Egypt's past.

Henry Salt, as British consul, was easily able to obtain a digging permit. He hired Belzoni to bring back the head of a giant statue from Luxor. He also told Belzoni to bring back any other relics that he thought had some value.

Belzoni and Sarah spent five and a half months preparing the enormous head for transportation to Cairo. They made several trips farther up the Nile and found many other artifacts, only a small number of which they were able to take back with them. Still,

Salt was pleased with their find, and he gave Belzoni two lion-headed statues of the god Sekhmet as a bonus, along with six months' salary.

With the success of his first trip, Belzoni now considered himself a true archaeologist. He thus decided to make a second trip up the Nile, to retrieve many of the artifacts he had been forced to leave on his first expedition. Sarah hadn't liked the squalid living conditions in Luxor, so she stayed in Cairo. Belzoni and a ragtag group of like-minded grave robbers traveled up the river in 1816. When Belzoni arrived in Luxor, he discovered that a Frenchman had hired all of the available workers in town to unearth the Temple of Karnak on the east bank of the Nile. So Belzoni went hunting for relics on the west bank. After some minor successes, he traveled to the Valley of the Kings, where he uncovered the spectacular Temple of Abu Simbel. It is famous for the four huge guardian statues at its entrance, each colossus more than sixty feet tall. But despite the beauty of the temple, there was little inside it that Belzoni could steal, so he returned to the Valley of the Kings.[5]

After discovering several minor tombs by applying his knowledge of engineering as to their possible locations, Belzoni made one of the most spectacular finds in the history of archaeology. He located the burial vault of Seti I, which was decorated with incredible wall carvings that still retained their original paint. Plus, there was an alabaster sarcophagus inscribed with hieroglyphics. Still, although finding the tomb was a major triumph, it didn't make Belzoni rich.

While in Cairo, Belzoni took some British tourists on a sightseeing trip to the Great Pyramid. Looking at the nearby pyramid of Khafre, whose entrance had never been discovered, Belzoni was struck with an idea about how to find a way inside. His hunch paid off, but, as was so often the case, the pyramid had been looted by grave robbers hundreds of years earlier.

Soon afterward, Belzoni had a falling-out with Henry Salt. Belzoni had assumed that all the relics he had found for the British consul were going to the British Museum. Instead, he learned that Salt was claiming ownership of the pieces and selling them to museums throughout Europe. What angered Belzoni the most was that the antiquities were being listed as from Salt's collection, and nowhere was there any mention that Belzoni was the person who had actually found them. Belzoni was not only not rich, he was also not famous. He returned to London, where he wrote a book about his adventures in Egypt, but he soon grew bored with city life. The last few years of his life he spent looking for new worlds to conquer.

The greatest grave robber of all time died of dysentery in December 1823, while searching for the source of the Niger River. Belzoni was buried beneath a tree in the town of Guato, in the kingdom of Benin. Decades later, British explorer Richard Burton looked for Belzoni's grave but could find nothing.

It's politically correct to protest the work of grave robbers like Belzoni and Salt, but the truth of the matter is that during the years these proto-archaeologists worked, the people of Egypt were not in the least interested in the heritage of their ancestors. Mummies were routinely ground up into powder to make all sorts of elixirs that supposedly cured a variety of ills. The tombs of the greats and not-so-greats of ancient Egypt were scoured for precious metals, and by the time of Belzoni, nearly everything worth taking had been gone for centuries. If it hadn't been for the efforts of Belzoni and Salt and the rest of their generation, most of the existing treasures of the pharaohs would have been smashed into dust for making concrete. Sometimes, greed works for the common good. But not always—as in the case of the Spanish conquistadors.

Growing up in the American Southwest, Indiana Jones was no doubt familiar with the Spanish conquistadors. The word *conquistador* means "conqueror" in Spanish, and that's primarily what the early soldiers and adventurers of fifteenth-, sixteenth-, and seventeenth-century Spain did—they conquered much of North and South America for their home country. Unlike the explorers for other countries, the Spanish conquistadors were less interested in establishing long-term colonies in their conquered land; their main concern was how much gold they could steal from the native populations. When the flow of gold ceased, within decades Spain slipped from being one of the major world powers to an also-ran.

The first Spanish invasion of America was the conquest of the island of Hispaniola (now divided into the nations of Haiti and the

Dominican Republic). Using Hispaniola as a base, Juan Ponce de Leon conquered Puerto Rico while Diego Velázquez invaded Cuba. The first settlement on the continent was Darien, in Panama, established by Vasco Núñez de Balboa in 1512. None of these new conquests provided any gold or rare spices, so they were considered failures by the Spanish government.

The conquistadors' first success was the conquest of the Aztec Empire by Hernando Cortes. The Aztec Empire covered much of what we now know as Mexico.

Francisco Pizarro was the conquistador who conquered the Incan Empire in South America. Pizarro was later appointed governor of all the lands he had seized for Spain. He was a harsh ruler, whose men despised him. In a famous death scene, Pizarro was trapped in his luxurious bedroom by a troop of his private guards. Bent with rage, Pizarro raised his sword and declared, "Who dies first?" This incredible act of courage was stolen by fiction writers over the years for many adventure stories. It was used to especially good effect by Robert E. Howard, the creator of Conan the Barbarian, in the first Conan story, "The Phoenix on the Sword," published in *Weird Tales* magazine in December 1932.

Although the conquistadors usually were greatly outnumbered by the Indian armies they fought, they were armed with several secret weapons, a number of which they were not even aware they possessed. The Spaniards carried smallpox and other deadly plague germs on their bodies. The diseases were completely unknown in the Americas, so the Indians had no natural antibodies in their blood to fight the diseases. Wave after wave of deadly outbreaks of smallpox and diphtheria swept through the Indian tribes of North, South, and Central America in the years after the conquistadors arrived. It's estimated that tens of thousands of natives died due to disease.

One conquistador whom Indy was sure to have studied in school was Francisco Vázquez de Coronado. Born in Salamanca, Spain, Coronado was made the governor of Nueva Galicia in New Spain (which today consists of the Mexican states of Jalisco, Nayarit, and Sinaloa). When one of the few survivors returned to Nueva

Galicia from a scouting mission, he told the governor about a city of gold called Cibola, which stood on a high hill, was built entirely of gold, and was not far from the Pacific Ocean.[6]

Excited by this report, Coronado assembled an expedition to find and claim the gold and other riches. He sent a group of men by sea to carry most of the supplies for the expedition and sent a second group to travel through the desert following the trail to the golden city. Coronado and Viceroy Antonio de Mendoza invested a lot of their own money in the mission. Mendoza, a close friend of Coronado, appointed him the commander of the expedition, with the goal of finding Cibola.

Coronado set out in early 1540, leading a huge expedition of more than three hundred Spaniards and thirteen hundred natives. Also accompanying the expeditionary force was Marcos de Niza, the newly appointed provincial superior of the Franciscan order in the New World. The one thing the Spaniards had learned was that whenever they explored new territory, they should bring a priest to try to convert the natives. Along with being greedy, the conquistadors were sanctimonious.[7]

Coronado spent two years hunting the city of gold in the deserts of the Southwest. During his trip, he went through parts of Arizona, New Mexico, Texas, Oklahoma, and even Kansas. At one point, the governor took ill and, while recuperating, sent some of his men on scouting trips. It was one such group that first discovered the Grand Canyon. None of these scouts ever made it as far north as Utah. Needless to say, Coronado never found any gold or fabled cities.

In 1542, Coronado was forced to return with his men to Mexico in an attempt to quell a native rebellion. Following pretty much the same route as he had followed from Mexico, Coronado didn't return to the country until several months after the rebellion was over. The long, fruitless treasure hunt had bankrupted Coronado, but he still remained governor of New Galicia until 1544. Two years after his trip, he retired from his government post and moved to Mexico City. He died there ten years later, in 1554. Thus ended the age of the Spanish conquistadors, not with a bang but with a whimper.[8]

The Cross of Coronado is the relic that Indy steals from the grave robbers when they dig up this ornamental crucifix in the Utah desert. According to the men who found the treasure, it was a gift given by Pizarro to Coronado in 1520. It's the same relic that Indy rescues twenty-five years later off the coast of Portugal and donates to Marcus Brody's museum. The only problem with this tale of finding the cross is that neither Coronado nor his men had ever visited Utah. How the relic made it to a secret location inside a mountain cave is, as they say in the movies, another story. The Cross of Coronado, like many of the antiquities in the Indiana Jones series, is an imaginary relic.

Circus Trains and Wild Animals

After stealing the cross, Indy leads his pursuers on a chase on top of a moving train containing the Dunn & Duffy circus. Unfortunately, the roofs of the cars, especially those transporting animals—including rhinos, snakes, and lions—aren't strong enough to support Indy's weight, and he falls into one where he encounters a lion. Indy grabs the nearest weapon he can find, a bullwhip, and snaps the whip just in time to send the lion scurrying back into the corner. In handling the whip for the first time, however, Indy also manages to cut his chin, thus creating his trademark scar.

After a perilous encounter with a rhino and a terrifying moment with snakes, Indy is able to escape the dangers of the train by using a trap door in the magician's car. His escape only postpones his inevitable capture, though, and he is forced to hand over the cross to the sheriff.

Normally, bullwhips are used by cowboys and farmers to control livestock. At its tip, a bullwhip can travel faster than the speed of sound, creating a sonic boom. This is due to the bullwhip's design and flexibility, as well as its length. Supposedly, bullwhip experts are so skilled that the whips rarely, if ever, touch an animal.

No one really knows for sure how the bullwhip originated. Some historians suggest that it was invented in Spain. Other experts believe it was created in South America. It seems pretty obvious to us, however, that even the ancient Romans used whips—remember the movie *Ben Hur*? Regardless of their origin, the whips have always been used the same way: to make people and animals snap to attention, run, and do what the whip holder wants.

Bullwhips are made in various ways, they come in different lengths, and they vary in cost. But basically, all bullwhips have a handle and a thong. The handle is somewhere between eight and twelve inches long. The braided thong is made from leather. Bullwhips are typically up to twenty feet long, though sometimes they are even longer. We cover bullwhips in greater depth in our earlier section about *Raiders of the Lost Ark*.

In the early days of the twentieth century, a circus was a traveling group of performers that normally included acrobats, clowns, and other novelty acts, along with a menagerie of wild animals that were made to perform tricks with an animal trainer. At the time, circuses were held under a big tent, and they criss-crossed the country by rail. Most historians believe that the circus started in ancient Rome, with chariot races in open-air stadiums. Philip Astley, an Englishman, and Antonio Franconi, a Frenchman, were considered the originators of the modern circus. They organized traveling circuses in Europe and England at the end of the eighteenth century.

Circuses were a popular form of entertainment in the United States in the years before the Civil War. Joshua Purdy Brown was credited as being the first circus owner to have his show perform under a canvas tent. P. T. Barnum introduced sideshows to circuses, while William Cameron Coup was the first circus promoter to use trains to transport a circus from town to town. Early circuses merely featured displays of exotic animals, but within a few years, equestrian acts became part of the show.

In 1833, Isaac Van Amburgh took animal acts to the next level when he entered a small cage containing a tiger, a leopard, and a lion. Circus historians consider Van Amburgh to be the first wild animal trainer. He soon became one of the most famous circus performers of the nineteenth century and even entertained Queen Victoria. As a showman, Van Amburgh liked to emphasize "the incredible danger he faced and the magnificent courage it took to face the beast in the cage. Dressed like a Roman gladiator, Van Amburgh forced his animals to perform tricks by cruelly beating them into submission with a crowbar."[9] One of Van Amburgh's most famous tricks was sticking his head inside a lion's mouth.

This new style of circus entertainment, wild-animal taming, became so popular that every traveling circus in the United States had to include a lion-tamer act as part of the show. Though these acts were all dubbed "lion taming," many of the shows featured other big cats, including panthers, cougars, and tigers. By the early 1920s, there were dozens of wild-animal acts touring with circuses around the country. Perhaps the most notable such act was that of Terrell Jacobs and his wife, Dolly, who performed inside a fifty-foot cage with fifty wild jungle cats. Another famous wild animal trainer was Clyde Beatty, who was credited as being the first man to use a chair to defend himself against lions and tigers.

By Indy's time, traveling circuses were one of the main forms of entertainment in rural America. These were the days before radio and TV, and motion pictures were still a novelty. When a circus train came to town, everyone took a break to attend the big show. In the early twentieth century, more than one hundred circuses performed around the country for audiences up to twelve thousand people. The strange, almost sinister, appeal of these traveling circuses was best captured by Ray Bradbury in his novel *Something Wicked This Way Comes*.

The Dunn & Duffy circus in the movie was fictional. Rhinos did appear in circuses, although their act usually consisted of their being walked around a circus ring. A rhino cannot be taught to perform tricks. Rhinos were primarily menagerie animals in circuses, and a rhino with a long horn would never be allowed in a circus due to the possibility of a performer or a sightseer being injured. While most circuses had a snake charmer as one of their acts, no circus on record has stocked hundreds of snakes. Snake charmers worked with big snakes, usually pythons or cobras. Moreover, many circus snake charmers removed the fangs from their snakes to ensure that no fatal accidents occurred during the performances. Although Indiana's fall into the snake car on the circus train made good theater, the snake car itself was a total fabrication.

The Search for Facts, Not Truth

It is now 1938, and Indiana Jones is engaged in another fight for the Cross of Coronado, this time off the coast of Portugal. The results are different this time, and within a few days our hero is back teaching a class in college, after which he gives the relic to his friend Marcus Brody.

It's in Indy's lecture to his college class that he remarks, "Archaeology is the search for facts, not truth. Truth is down the hall in the philosophy class."

Such beliefs, that the study of past cultures and societies is dependent on physical objects and not on theoretical concepts, might have been considered true in the first half of this century, but it has little, if any, validity today. Modern archaeology is the study of human society as it developed over the ages. While archaeologists are still interested in facts, theories are equally, if not more, important.

Not that Professor Jones's statement about facts, not truth, makes much sense when we recall Indy's previous adventures involving the Lost Ark and the Temple of Doom. In both cases, facts are trumped by fantasy. Facts didn't and couldn't explain what happened when the Ark of the Covenant was opened. Nor could mere facts explain the horrid talent that enabled Mola Ram to extract the beating heart from the chest cavity of his still-living victim.

Indiana's remarks also don't make much sense when contrasted with his later statement that "archaeology isn't the hunt for lost cities buried in the sand." Surely, the Germans who were searching for the Lost Ark would have disagreed with that statement, as Indiana would have just a few years earlier, when he succeeded in finding the Ark buried beneath the sands that covered the ancient city of Tanis.

As the authors of this book have both taught college classes, they understand that despite the best of motives, not every word said in a classroom is the absolute truth. Professor Jones's definition of archaeology seems to have shifted every time his mood changed.

Indy's assignment to his class to study Flinders Petrie's discovery of Naucratis in 1885 was only slightly off and could easily be attributed to a slight memory slip. Flinders Petrie was a famous

British archaeologist, and one of his earliest finds—in 1884, not 1885—was Naucratis, an ancient Greek settlement in Egypt located on the eastern canal bank about ten miles west of the Rosetta branch of the Nile. Although the discovery of Naucratis was eclipsed by many other, more important discoveries made in Egypt by Petrie, in 1938 Petrie's early work would still have been taught in most college archaeology classes. But, surely, Indy would have known the proper year of Petrie's first major discovery.

In a conversation with Walter Donovan, Indiana describes his father as the teacher of medieval literature whom students hated to get. While that might have been true of Dr. Henry Jones in his later years, it's hard to believe that a medieval literature professor lived in the wilds of Utah in 1912. Or that an elderly medieval literature professor was capable of conducting a hunt throughout Europe and the Middle East for the Holy Grail. Since Indy describes archaeology as being "70 percent of the time spent in the library," it seemed much more likely that his father was one of those archaeologists who devoted most of his time to doing research on paper, not in the field. He was a man who was much like Indy's old friend Marcus Brody.

That Indiana Jones was the son of a famous archaeologist was unusual but not unique. Emile and Paul Vouga were father and son Swiss archaeologists who separately investigated the Iron Age site of La Tène. Emile Vouga excavated the site in the mid-1870s, working with William Wavre. His son, Paul Vouga, worked at the same site from 1907 to 1918, also with Wavre.

George and David Stuart, another father and son duo, have studied the Mayan people. George Edwin Stuart is an archaeologist and a cartographer. He earned a BS in geology from the University of South Carolina in 1956, an MA in anthropology from George Washington University in 1970, and a PhD in anthropology from the University of North Carolina at Chapel Hill in 1975. Beginning in 1958, he focused on Mayan and Mesoamerican archaeology, and he also performed excavations in the Yucatan and Quintana Roo, Mexico. His son, David Stuart, is an expert on the ancient Mayan people, in particular their written language, and he serves as professor of Mesoamerican Art and Writing at the University of Texas at Austin.

Taking family science to an entirely new level, famous anthropologist Mary Leakey was married to Louis Leakey, also a famous anthropologist, and was the mother of Richard Leakey, yet another famous anthropologist. Mary was best known for her excavations at Olduvai Gorge in Africa, where she discovered the Zinjathropus Man, which proved not only that humanity's ancestors originated in Africa, but that human development stretched much farther back in time than anyone had imagined. Mary also discovered the fossil footprints at Laetoli, in Tanzania. Preserved in volcanic ash, these footprints were the earliest record of bipedal gait, some three million years old. Her husband, Louis, was a noted archaeologist and anthropologist. A religious man, he was famous for stating, "Nothing I've ever found has contradicted the Bible. It's people with their finite minds who misread the Bible."[10] Their son, Richard Leakey, is a well-known anthropologist, ecologist, and author. He is a visiting professor at the State University of New York at Stony Brook.

The Jones family was thus neither unique nor unusual in having two archaeologists in the family. Nor was it particularly odd that they both taught at the same university.

Barnett College, identified in *The Last Crusade* and sometimes called Marshall College, as listed on the Official Indiana Jones Web site, was where both Indiana and his father taught during the late 1930s. According to mail glimpsed on the elder Professor Jones's desk, the school was located somewhere in the vicinity of Ferndale, New York. However, Ferndale, located in the Catskill Mountains, some distance north of New York City, was not home to any college or university during the 1930s (and is not today). The invention of industrious screenwriters, the pastoral college campus with its archaeology department and museum of antiquities was most likely based on a somewhat romanticized notion of Yale University, located in New Haven, Connecticut.

Walter Donovan, American Nazi

Indy is taken to New York City, where he meets Walter Donovan, a rich industrialist engaged in a search for the Holy Grail. Donovan believes that the Grail will grant immortality to whoever owns it. He believes

a clue to the Grail's location is located somewhere in Venice. Indy's
father went hunting for that clue and has disappeared. When Donovan
lost faith in Dr. Jones's loyalty, he arranged for the professor to be kid-
napped near an important spot in a library. Donovan gambles on Indy
finding the Grail that his father was not able to. A ruthless, extremely
wealthy man, Donovan isn't going to let patriotism stand in his way.

It doesn't take much to convince Indiana to go searching for his father
and the Holy Grail.

There were two villains in *Indiana Jones and the Last Crusade*. The
first, and more obvious, villain was the German war machine, as
personified by Dr. Elsa Schneider and SS colonel Vogel. The second menace faced by Indy was rich American industrialist Walter
Donovan, who was cooperating with the Nazis in hopes of finding
the Holy Grail. Donovan was less a traitor than he was a practical
man. He wanted to live forever and was willing to throw his support
behind the side he thought would locate the Grail, the Nazis.
Donovan, it turned out, was the man who'd hired the senior
Dr. Jones in New York and persuaded him to lead the Grail quest
in Venice.

Sad but true, men like Walter Donovan existed in pre–World
War II America. There was a great deal of opposition to the United
States entering the war on the side of the Allies, and many people
felt that Franklin D. Roosevelt was a pawn of the "Jewish Bankers
Conspiracy." For a number of years, an American Bund, a branch
of the Nazi Party, existed in New York City. It's quite possible that
the men who brought Indy to see Donovan were members of the
German-American Bund. Before the war broke out, it wasn't a
crime to be a Nazi and support Hitler's racial agenda—even if you
were rich and famous.

In the early 1930s, the neo-Nazi German American Bund was
created from two established organizations: the National Socialist
German Workers Party and the Free Society of Teutonia. National
Socialist German Workers Party member Heinz Sponknobel
brought the two organizations under the umbrella of the Friends of
New Germany with the supposed objective of fostering peace between Germany and the United States. Of course, the Party was

the Nazi Party, and the Free Society of Teutonia was a militant organization that supplied strong Nazi support, so the Bund was not
spawned from peaceful roots.

Shortly after the formation of the Friends organization, it came
under fire from two different fronts. Friends of New Germany was
attacked by the Jewish residents of New York City, who organized a
boycott of German goods in the German neighborhood of Yorkville on Manhattan's Upper East Side. Plus, Jewish congressman
Samuel Dickstein, a Democrat who represented New York City,
decided to investigate the group's ties with Germany.

The Friends of New Germany countered the boycott with wide
dissemination of propaganda that tried to show that the Jews had
made up and lied about everything that the Germans had done to
them. Meanwhile, several factions tried to take control of the
Friends. In 1934, Sponknobel was removed as the group's leader,
while the Dickstein investigation concluded that the Friends of
New Germany was a front for the Nazi Party of the United States.

Adolf Hitler, realizing that the Friends were drawing unwanted
attention to German activities in the United States, mandated that
Germans withdraw from the group. In an attempt to make the
organization seem like nothing more than a grassroots German-
American social organization, Hitler appointed naturalized American citizen Fritz Kuhn as head of the Friends. The name of the
organization was changed to the German-American Bund.

Kuhn was born in Munich, Germany, in 1896 and served in
World War I as a machine gunner in the German Army. He joined
the Nazi Party in 1921, got a degree in chemical engineering, then
moved to the United States in the late 1920s. He became a U.S.
citizen in 1934. Kuhn spoke terrible English but was a dynamic
speaker and leader. He took several competing German-American
organizations and forged them into one larger group, the German-
American Bund.[11]

A big, powerfully built man, Kuhn believed that Hitler was
Germany's savior and that he, Kuhn, was America's führer. Shortly
after organizing the Bund, Kuhn started a Camp Siegfried on Long
Island, where German Americans could dress up in black leather
boots and train for "der Tag," the Day when they would rule the

United States. Children in uniform would march in formation past their führer, raising their arms in the palm-down Hitler salute, while singing, "Youth, youth, we are future soldiers." Similar camps were opened in other states: Efdende North in Michigan, Deutschhorst in Pennsylvania, Nordland in New Jersey, and Hindenburg in Wisconsin.[12]

The Bund held beer festivals, organized lectures about Nazi music and art, and sponsored many sports teams and other activities. Wearing Nazi uniforms, they marched in many parades, openly in honor of Hitler. In addition, they distributed in front of Jewish-owned establishments leaflets that read "Patronize Aryan Stores."

The German-American Bund members paid for the fairly substantial expenses of all of these activities; they bought the uniforms with Nazi insignias, the rings, the emblems, and the membership certificates. Advertising in the Bund's various publications wasn't cheap, either. The money went to the national headquarters of the group in Yorkville, where leader Kuhn treated it as if it was his personal war chest. Bookkeeping was virtually nonexistent, a mistake that would cost him dearly in the years ahead. Like many men riding the crest of a wave of popularity, the American führer grew overconfident. Hubris, the Greeks called it.[13]

In late 1936, Kuhn and a group of his wealthiest supporters attended the Olympics in Berlin, Germany. At the games, Hitler met with many people who wanted photos taken with him. Kuhn got to meet his idol for only a few moments, but it was enough that he had his photo taken with the real Führer. The picture of the two men was printed in newspapers all across the United States, giving Kuhn brief fame. In conversations with his supporters afterward, Kuhn claimed that Hitler agreed that Kuhn was to be the American führer.

Back in the United States, Kuhn had a series of extramarital affairs and was often seen with flashy women in nightclubs. He enjoyed wearing his Nazi uniform in public and capitalized on his "relationship" with chancellor Hitler. Kuhn was famous enough that a number of tabloid newspapers covered his every speech and scandal. He was considered the leading anti-Semite in the United

States, a title he was proud to proclaim. According to the newspapers, Kuhn had two hundred thousand armed Nazi men at his American command. In addition, the newspapers claimed that there were fifty thousand Nazis in Connecticut. Kuhn became so infamous that he even appeared in thinly disguised descriptions as the villain in a number of pulp magazine serials.[14]

Kuhn's greatest triumph and the biggest event ever held by the German-American Bund took place on February 19, 1939. The Bund organized a rally in Madison Square Garden in New York City. Shocking as it was, more than twenty thousand people attended. Three thousand men in Nazi uniforms led Kuhn to the speaker's platform. There he stood, and behind him was a huge picture of George Washington surrounded by American flags and Nazi swastikas. Kuhn called President Franklin Delano Roosevelt "Frank D. Rosenfeld," referred to the New Deal as the "Jew Deal," and made other incendiary claims, such as raging that a Jewish-Bolshevik conspiracy ruled America.[15] Fights erupted during the rally between Kuhn's storm troopers and Jewish World War I veterans.

Needless to say, Kuhn's activities finally drew the attention of the House Un-American Activities Committee. In 1939, the U.S. government stated that the German-American Bund was closely aligned with the German Nazi Party. Soon after that, a New York tax investigation concluded that Kuhn had embezzled $15,000 from the Nazi rally at Madison Square Garden. The German-American Bund refused to file criminal charges against Kuhn, but the district attorney of New York wasn't so forgiving. He prosecuted Kuhn, and the Bund leader was sent to Sing Sing. When the war ended, Kuhn was deported to Germany. He died mostly forgotten in 1951.

A number of prominent Americans in the 1930s were famous (or infamous) for their admiration of Hitler and their hatred of Jews. The most notable of these celebrities were Henry T. Ford and Charles Lindbergh.

Ford was notorious for his anti-Semitic views long before Hitler's rise to power. In 1920, a small newspaper he owned, the *Dearborn Independent*, published a supposedly true book of secret Jewish rituals that included killing Christian children as blood sacrifices. The book was titled *The Protocols of the Learned Elders of Zion*. In February 1921, in an interview with Ford published in the *New York World*, the automobile magnate said "[T]he only statement I care to make about the Protocols is that they fit in with what is going on."[16] Along with the *Protocols*, the *Dearborn Independent* published a series of anti-Semitic articles that were collected in four hardcover volumes titled *The International Jew, the World's Foremost Problem*.

During the early years of the Nazi movement, Hitler read the four books. Ford was the only American mentioned in Hitler's *Mein Kampf*. In 1938, the German consul in Cleveland gave Ford, on his seventy-fifth birthday, the Grand Cross of the German Eagle, the highest award given by Nazi Germany to a foreigner.

Another highly respected American who praised Hitler before World War II was Charles Lindbergh. He had become a national hero after flying solo across the Atlantic in 1927. Lindbergh was a staunch anticommunist, and he felt that Hitler's Germany served as a reliable bulwark against Europe's falling to the Russian threat. Unfortunately, Lindbergh was politically naive and, when he spoke, revealed an innate dislike of Jews. In a speech made in Des Moines, Iowa, at an antiwar rally on September 11, 1941, Lindbergh declared that the three groups that were pushing the United States toward war with Germany were "the British, the Jews, and the Roosevelt Administration."[17] In the same speech, he complained about the Jewish people's "large ownership and influence in our motion pictures, our press, our radio and our government."[18]

In a time before television and twenty-four-hour news stations, Lindbergh's remarks didn't result in his immediate downfall, but the reaction to his speech was extremely negative. Lindbergh was forced to defend his comments by stating again and again that he was not anti-Semitic, but his explanation didn't ring true, especially when journalists noted the flyer's previously expressed admiration for Hitler and his government.

Lindbergh served as a civilian consultant to several important aircraft companies during World War II. After the war, he wrote a number of books, including an account of his transatlantic flight, *The Spirit of St. Louis*, which won a Pulitzer Prize. He died in Maui, Hawaii, in August 1974.

Another prominent American who gained some measure of notoriety for his dealings with the Third Reich was U.S. senator Prescott Bush, the grandfather of George W. Bush. Prescott Bush worked for Brown Brothers Harriman (BBH), an investment company that was German industrialist Fritz Thyssen's U.S. base. Thyssen was famous for his role in helping to finance Hitler in the 1930s before their relationship petered out in 1940. Bush was the director of the New York–based Union Banking Corporation (UBC), which represented Thyssen's interests in the United States. Bush was also on the board of one of the companies that formed a multinational network of front companies that enabled Thyssen to move assets throughout the world.[19]

Even after the United States entered the war and there was significant information available about the Nazis' plans and policies, Prescott Bush continued to work for and profit from companies that were closely involved with German businesses that financed Hitler's rise to power.

At least one former U.S. Nazi war crimes prosecutor argued that Bush's actions should have been grounds for prosecution for giving aid and comfort to the enemy. It has also been suggested that the money Prescott Bush made from these dealings helped to establish the Bush family fortune and set up its political dynasty.

Another famous participant in U.S. politics who had an unsavory relationship with Nazi Germany was Joseph Patrick "Joe" Kennedy, U.S. businessman, political leader, and father of President John F. Kennedy and Senators Robert F. Kennedy and Ted Kennedy. A leading Democrat in the Irish Catholic community, Kennedy had nationwide business and political affiliations and friendships. In 1938, President Roosevelt appointed Joe Kennedy as ambassador to Great Britain. Kennedy rejected Winston Churchill's warning that compromise with Nazi Germany was impossible. Instead, Kennedy supported the policy of appeasement put forth by British prime minister Neville Chamberlain as the best way to prevent a war breaking out all over the world. As Nazi persecution of the Jews escalated in 1938, Kennedy attempted, without State Department approval, to obtain an audience with Adolf Hitler.[20] During this period, Kennedy maintained an extremely negative stance about Jews. According to Harvey Klemmer, who served as an embassy aide to Kennedy, the ambassador routinely called Jews "kikes" and "sheenies." Kennedy also stated, "Individual Jews are all right, Harvey, but as a race they stink. They spoil everything they touch."[21]

On June 13, 1938, Kennedy met with Herbert von Dirksen, the German ambassador in London, who reported to Berlin "that [Kennedy] himself fully understood our Jewish policy."[22]

Like most friends of the Third Reich in 1940, Kennedy feared that a third term for Roosevelt meant war. Laurence Lerner reported that Kennedy believed "that Roosevelt, Churchill, the Jews and their allies would manipulate America into approaching Armageddon."[23]

While none of these men ever cooperated with the Nazis on the same level as Walter Donovan, it's easy enough to see how they could serve as examples for Donovan—and given the temptation of obtaining eternal life from the Holy Grail, how they might have followed the same path.

The Holy Grail

According to Walter Donovan, the Grail was the cup used by Jesus Christ at the Last Supper. Later, the same cup caught the blood of Jesus when he was stabbed in the side by a spear wielded by a Roman centurion while Jesus was on the cross. Donovan also states that the cup was entrusted to Joseph of Arimathea after the Crucifixion and that it disappeared for more than a thousand years.

Somehow, Donovan continues, the Grail ended up with three brothers who were crusaders, one of whom was entrusted with keeping the Grail safe, while the other two went out into the world to tell other worthy men that the cup still existed. Of course, both of those brothers died without revealing the location of the Grail, and it remained hidden away with the third brother, unseen by anyone else for the next seven hundred years.

Donovan wants the Grail because he feels certain that drinking from the same cup used by Christ at the Last Supper will endow the user with immortality. He is also positive that water from the Grail will heal all wounds, even life-threatening ones.

What exactly is the Holy Grail, what are its powers, and where is it located (if that location is known)? Most important, how do we know these things?

The Holy Grail is a name that commonly refers to either the cup Jesus Christ used at the Last Supper or the dish used for the "Paschal lamb" at the same meal. The Paschal lamb was the lamb that the Children of Israel in Egypt were commanded to eat by Moses, following certain rituals as part of the celebration of Passover. The Last Supper, of course, was a Passover meal eaten by Jesus before he was taken prisoner by the soldiers of Pontius Pilate. In the earliest writings about the Grail, it is unclear whether it is a cup or a plate.

The word *grail* isn't easily explained either. The most common explanation of the meaning of the word comes from a thirteenth-century monk, Helinandus. He told of a Latin book titled *Gradale* that was written by a hermit about the plate used by Jesus at the Last Supper. According to Helinandus, the word *gradale* meant a wide and somewhat deep dish that held costly food that was served in small portions known as *gradatim*.[24] A later interpretation of the term *san gréal*, Old French for "holy grail," was derived by moving one letter to spell *sang réal*, or "kings blood."

The Grail legend is bound closely to the legends of King Arthur, as well as to the story about one of the Knights of the Round Table, Percival. Originally, all of these legends were independent of one another. Yet somehow they became so intertwined that it's impossible to tell which came first and which, if any, are based on true happenings.

The truly important aspects of the Grail legend were developed in France and England between the end of the twelfth century and the first forty or so years of the thirteenth century. After 1350, no major elements were added to the story. While most of the legends were written in French, versions also existed in English, German, and Italian. A few Grail stories were even written in Norwegian and Portuguese. Some adventures were translations of the French epics, but a number of them were new tales created by storytellers from those countries. Two major themes dominated the Grail myths. The first dealt with a hero searching for the Holy Grail and his exploits during that quest. The other theme described in detail the history of the Grail.

The earliest Grail romance was the "Conte del Graal" of Chrétien de Troyes and many others. This was a huge poem of more than fifty thousand verses that was written between 1180 and 1240. More than any other work, it described the story of the Grail and its history and the details of the Grail quest involving the knights of Arthur. Of the "early history" versions, the oldest is the trilogy composed by Robert de Boron, written between 1170 and 1212. Unfortunately, only the first part and a portion of the second remain. The most detailed history of the Grail is contained in the *Grand St. Graal*, a lengthy French prose romance composed in the first half of the thirteenth century. Several French prose romances from the thirteenth century devote some space to the history of the Grail. One of these, *Queste del St. Graal*, or *Graal*, was well-known to English readers because it was embodied nearly complete in Malory's *Le Morte d'Arthur*.[25]

Most scholars consider Chrétien's poem to be the oldest Holy Grail story. The poem tells of a knight named Perceval who visits a mysterious castle. In the main chamber of the edifice, a jeweled bowl is brought into the room by a beautiful maiden. Along with the Grail, she carries a wide silver platter and a blood-covered spear. The other knights in the room all bow before the Grail. Of course, Perceval does not learn anything about the mysterious Grail. Without doubt, Chrétien planned to cover that in a follow-up epic, but he died before he could write the next story. For the next hundred years, other writers continued the story, seeking to capitalize on his fame.[26]

In Chrétien's original story, the Grail had no religious connotations. Over time, as other writers continued the saga and retold the original, the Grail became a religious artifact of great importance. Supposedly, Perceval learns that the Grail was the actual bowl from which Jesus and his disciples ate the Passover meal. That same container was later claimed by Joseph of Arimathea. When Christ was removed from the cross, Joseph collected Christ's blood in the bowl. In later stories, the Grail somehow became one with the Chalice of the Eucharist. The spear in the story, of course, became the fabled lance that Roman soldier Longinus thrust into Jesus.[27]

Chrétien made it clear that the Grail in his poem was some type of ornamental bowl or a jeweled serving dish. Other authors, however, took great latitude with Chrétien's version. They decided to describe the Grail in other ways and with other meanings. In the epic poem *Parzival*, Wolfram von Eschenbach declared that the Grail was a stone that had dropped from heaven. According to Eschenbach, the Grail had protected all of the angels who had remained neutral during Satan's revolt against God. Robert de Boron wrote that the Grail was the cup used at the Last Supper. In yet another version of the story, Perceval was presented with a platter holding the head of his brother (shades of *Salome!*). In *Le Morte d'Arthur* by Sir Thomas Mallory, which took many of the Grail legends and tied them into the story of King Arthur, the Grail was the object sought after the by the Knights of the Round Table.[28]

The Holy Grail is a mainstay of modern Western mythology. It is a common mystical object in legends throughout Great Britain, France, and the United States partly because of its religious significance. Equally important, though, is the tie-in the Grail has with the legends of King Arthur. The stories of Arthur, Lancelot, and Galahad and the Grail quest have been told repeatedly, in everything from movies to novels to comic books. Based on the popularity of the Arthur legend, most Americans and Brits assume that the Grail is famous throughout the world. That's not the case, as there are no stories about the Holy Grail in the folk tales of Eastern Europe. The Eastern Orthodox branch of Christianity has no traditions concerning the blood of Jesus. Nor are the myths of the Grail popular in predominantly Catholic Spain. Latin America is another area where the legend has never taken hold, and King Arthur and his noble knights are little more than characters in movies and television shows. In many ways, Indiana Jones's adventure in search of the Holy Grail might have been the first exposure many people ever had with the legend.[29]

According to the New Testament, Joseph of Arimathea was a wealthy inhabitant of the Biblical town, who donated his tomb for Jesus to be buried in after the crucifixion. In Mark 15:43, Joseph was described as an "honorable counselor, who waited for the kingdom of God." Most scholars felt that the word *counselor* indicates that Joseph was a member of the Sanhedrin, the ruling body of Jewish lawmakers. That he "waited for the kingdom of God," implied that Joseph was a believer in Christ's divinity. This notion is confirmed by the passage John 19:38, where he states that Joseph was secretly a disciple of Jesus.

The same passage in John told how, after the death of Jesus, Joseph went to Pontius Pilate and asked that he be given Christ's body so that Jesus could have a decent burial. With the governor's permission, Joseph went to Golgotha, and with aid from Nicodemus, removed Christ's body from the cross. According to Luke 23:53 and 23:55, the men brought the body to a tomb that Joseph had made for himself on his estate. In the presence of some women including Jesus' mother, Mary, and Mary Magdalene, they placed the corpse in the tomb.

It's worth noting that according to biblical scholars, very few people or events are discussed by all four evangelists. If a topic is covered in Matthew and Mark, it is not covered in John or Luke. Or if something is mentioned by Luke, it is not in Mark. However, Joseph of Arimathea and his involvement with Christ's burial is discussed by all four writers of the Gospels. Perhaps for this reason, Joseph is considered a saint by the Catholic, Lutheran, and Eastern Orthodox Churches. His feast-day is March 17 among Catholics. It is July 31 in the Eastern Orthodox Church as well as in the Lutheran Church.[30]

As you might imagine, Joseph of Arimathea figures prominently in the Grail legends of England. He is first mentioned by Robert de Boron in *Joseph d'Arimathie*, which was written in the late twelfth century. In that story, Joseph is given the Holy Grail by Jesus Christ himself, who appears to Joseph in a vision. Christ then tells Joseph and his followers to take the Grail to England, which of course, they do. According to de Boron, the Grail has mystical powers, including the ability to heal terrible wounds. This notion later became an integral part of the King Arthur legend. Many subsequent authors embellished the Grail story by having Joseph catch the blood of Christ as it dripped from his body during the burial. Still other writers described how once Joseph reached England, he established an order of guardian knights to protect the Grail and keep it hidden forever. Over the centuries, this story merged with the legends of King Arthur and the Knights of the Round Table. More recently, the Grail legend became part of the modern myths circulating about the Knights Templar.

To us, the story of a repentant politician who made his way from the Middle East to Great Britain during the first century seems outrageous and totally unbelievable. But it evidently rang true with many British historians over the centuries. In many Grail legends, Joseph of Arimathea is described as the first bishop of Christianity. John of Glastonbury, who wrote in 1350 the history of Glastonbury Abbey, told how Joseph traveled to England, bringing with him the wood cup used by Jesus at the Last Supper. Other legends, equally dramatic and embellished, have become a part of British folklore. Needless to say, more than one American author has claimed that the Grail now resides in the United States, having made its way here from England via the machinations of one secret society or another.

Over the centuries, a number of chalices have been proclaimed to be the actual Grail. This makes sense, of course, because Grail authors insist that the cup resides in many different countries. Unlike the cup in *The Last Crusade*, these chalices have usually not been the simple wood cup of a carpenter. They are typically quite ornate and made from precious jewels and gold. It is thought, for example, that a holy chalice in a cathedral in Spain was transported to Rome by Saint Peter in the first century, and two hundred years later was brought to Spain by Saint Lawrence. As best as can be ascertained by archaeologists, the chalice is a Middle Eastern cup ranging back to the first century. They believe that the cup's origin was in Antioch, Syria, and often call the cup the Antioch Chalice. The base and stem of this holy chalice consists of gold and gemstones, and the cup has even served as the official papal chalice. Underscoring the belief in the religious import of the relic, Pope Benedict XVI used the holy chalice as recently as 2006.[31]

The Crusaders in the Holy Land found a gorgeous emerald cup that they believed was the Grail. This cup was brought to Italy and named the Chalice of Genoa. When Napoleon conquered Genoa, he shipped the chalice to Paris, but it broke en route. The beautiful emerald cup was actually nothing more than green glass.[32]

Legends say that the Grail is buried beneath Rosslyn Chapel in Scotland, or that it lies deep in the spring at Glastonbury Tor. Other stories claim that a secret line of hereditary protectors keep the Grail. Still others maintain that the Grail was hidden by the Knights Templar on Oak Island, in Nova Scotia's famous "Money Pit." Local folklore in Accokeek, Maryland, says that it was brought to the town by a closeted priest aboard Captain John Smith's ship.

The Grail has been used in numerous works of fantasy and science fiction. Marion Zimmer Bradley's *The Mists of Avalon* has the Grail as one of four objects that symbolize the four elements: the Grail itself (water), the sword Excalibur (air), a dish (earth), and a spear or a wand (fire). Michael Moorcock's novel *The War Hound and the World's Pain* tells the story of a Grail quest set in the era of the Thirty Years' War.

The Holy Grail is one of the most popular topics in occult nonfiction. Since such books on the Grail do not have to prove any theories and because the past history of the Grail is so muddled, books linking the Grail with the Knights Templar, the Shroud of Turin, and every major religious controversy of the last thousand years are commonplace. While the Grail's powers are never spelled out in any of the early works written about it, that hasn't stopped so-called nonfiction authors from giving it powers that range from bestowing immortality on anyone who drinks from it to changing water into wine. Because no one knows where the Grail is or even exactly what it is, no claim about it is too outlandish or too strange.

One of the most famous nonfiction books on the Grail legend is *The Sign and the Seal* by Graham Hancock. In his work, Hancock links the story of the Holy Grail with that of the Ark of the Covenant, which he claims still exists and is kept hidden by an offshoot Jewish cult in Ethiopia.

No more believable, but evidently believed by many, is *Holy Blood, Holy Grail* by Michael Baigent, Richard Leigh, and Henry Lincoln, who claim that Jesus did not die on the cross but instead lived on and married Mary Magdalene. Their children and descendants formed the Merovingian bloodline, which continues to this day. In *Holy Blood*, the "Grail" merely means Mary as the receptacle of Jesus' bloodline. The book is filled with long, involved arguments about secret societies, religious orders, and mysterious events that, not surprisingly, cannot be verified by ordinary research.

Using similar ideas, Dan Brown's best-selling novel *The Da Vinci Code* forwards the same concept. Brown identifies the Grail as Mary Magdelene and mentions that she was once buried beneath Rosslyn Chapel in Scotland. According to Brown's novel, she now rests beneath the floor of the Inverted Pyramid entrance to the Louvre. Needless to say, Brown's assertions didn't thrill the French government, because it suddenly found the Louvre assailed by a horde of curious treasure seekers.

A number of historians consider early Celtic mythological stories about a fabulous cauldron to be the source of the idea that the Grail possesses mystic powers to heal the wounded and grant everlasting life. Over the centuries, according to scholars, these stories became intertwined, and as Christianity replaced paganism in Britain, so did the Grail replace the cauldron, gaining all of its powers in the exchange. No one, neither scientist, priest, nor religious scholar, has a solid explanation as to why drinking liquid from the Holy Grail would grant anyone eternal life.

The legend of the three knights' quest for the Holy Grail was rather mangled by the Indiana Jones series. As part of the tales of the Round Table, the quest for the Holy Grail was one of the last and most important missions undertaken by the knights of King Arthur. A vision of the Grail prompted all of the knights to seek the Grail, but after many adventures, only three, Sir Galahad, Sir Bors, and Sir Percival proved worthy. Galahad was the one to retrieve the Grail, and, after doing so, he died. The Grail did heal deadly wounds, but drinking from the cup did not grant immortality. The three knights were definitely not brothers, and none of the three were crusaders. Indy's research might have fit well as part of the story of the *Last Crusade*, but it wasn't based on anything faintly resembling a Grail legend or history.

The Rising Nazi Menace

It's the late 1930s, and Indiana is well aware that Adolf Hitler is obsessed with the occult. He knows that the Nazis are most likely searching for the Grail and that his father is in danger from the Führer's minions.

After his encounters with the Nazis in 1936 in *Raiders of the Lost Ark*, Indiana Jones knew that the German storm troopers would do anything to satisfy their leader's wishes—especially when it concerned some occult item that would give the Nazis incredible supernatural powers. Thus, it was no surprise that the Nazis were anxious to find the Holy Grail. With a cup that could grant the user eternal life, Hitler's henchmen realized that it was in their power to create an empire, a Third Reich, that literally could last a thousand years. A thousand years all ruled by the same man, Adolf Hitler.

As discussed in part 1, the head Nazis, particularly Hitler and Himmler, were obsessed with the occult. Both men believed that it was their destiny to rule the world. The loss of the Ark of the Covenant to Indiana must have driven Hitler into paroxysms of rage. There's little question that the Führer wanted his men to do anything necessary to find the Holy Grail. And that his instructions to Elsa Schneider made it clear that once the Grail was found, there was no reason to keep the Jones men alive.

In 1936, around the time that *Raiders of the Lost Ark* took place, Nazi troops occupied the Rhineland, territory sacrificed as part of the Treaty of Versailles. Hitler secretly gave his troops orders to retreat if challenged by the French, but the French did nothing. The German economy geared up for war in four years. In 1937, a Gallup poll taken in the United States revealed that 94 percent of all Americans wanted to stay out of a war overseas.

In early 1938, Hitler demanded that Germans in Austria and Czechoslovakia be allowed to vote on whether they wanted their countries to become part of Germany. The chancellor of Austria called for a public referendum to be held on the question on March 13. On March 12, German troops marched into Austria and annexed the country. In England, Churchill made a speech saying that now was the time to stop Hitler. No one else seemed to agree. Representatives of Sweden, Denmark, Norway, Belgium, Luxemburg, and the Netherlands met to discuss how they could remain neutral in any conflict between the major powers.

In September, Neville Chamberlain, the British prime minister, met with Adolf Hitler and said that the Führer seemed to be a man who could be relied on to keep his word. At the end of September, Germany annexed the Sudetenland from Czechoslovakia. Nothing was done to stop him. The next day, Chamberlain declared that by signing the Munich agreement, which allowed Germany to seize part of the Czech Republic, Britain had acted honorably. "I believe it is peace for our time," said Chamberlain to a cheering crowd in London.

This was the world and the Nazi juggernaut that Indiana Jones faced in the fall of 1938.

The First Crusade

When Indiana Jones begins his search for the Holy Grail, his only clue is that the sacred goblet was found by three brothers, all of them knights, during the First Crusade. According to legend, the three brothers hid the Grail somewhere in the desert of what is now Turkey. One brother remained with the Grail, while the other two brothers left him and went to spread the word of its location. Indy must first find the tablets left by the two brothers that reveal the location of the church where their brother is waiting with the Grail.

So what were the First Crusades, where did they take place, and why were the three brothers involved? And how were the First Crusades related to the Holy Grail?

The Crusades were a series of military wars waged for religious reasons by Christians over the years 1095 to 1291. These wars were fought with the approval of the pope in the name of Christianity. The First Crusades were fought to recapture the city of Jerusalem and the surrounding Holy Land from the Muslims. These early Crusades were initiated as a response to a call for aid from the Eastern Orthodox Byzantine Empire against the Muslims who had gained control of Jerusalem and the Holy Land in 614.

Originally, in the seventh century and the hundreds of years that followed, there was little concern in Europe that Arabs held the land of Israel. Pilgrims were still allowed to visit Christian holy sites, and monasteries in the Holy Land were left alone to practice religion as they pleased. In Europe, the kings and the nobles were more concerned with the violent raids of the Vikings, as well as with invasions from the west by the Slavs and Magyars. That all changed early in the eleventh century.

During the early years of the eleventh century, a group of Turkish Muslims known as the Seljuks gained control of the Holy Land. Deeply religious, they killed many Christian pilgrims and clergymen. After a few years, the Seljuks came to the realization that the wealth of Jerusalem derived from the pilgrims, and they began to treat the pilgrims better. But they did impose a toll tax on all visitors to Jerusalem. Stories of the violence done to Christian pilgrims slowly made their way back to France and England and, instead of discouraging more pilgrimages, seemed to encourage such journeys.

In March 1095, the ruler of the Byzantine Empire, Alexius I, asked Pope Urban II for help in defending his empire against invading Muslim armies. In November of that year, Urban called the Council of Clermont to discuss the matter further. The pope, in his message to the faithful, told the bishops and the abbots to bring with them the prominent lords in their provinces.

The council lasted from November 18 to November 28, 1095, and was attended by about three hundred clerics and many nobles from all over France. On November 27, Urban spoke about the problems in the east, as he declared religious war against the Muslims who had occupied the Holy Land and were attacking the eastern Roman Empire. Urban gave an impassioned sermon in which he urged his audience to wrest control of Jerusalem from the hands of the Muslims. He spoke of the problems of nobles' violence against each other; the solution was to turn swords to God's own service. The pope promised remission of sins for anyone who would die in the undertaking. His speech provoked the crowd into wild enthusiasm, and they continually screamed, "*Deus lo volt!*" (It is God's will!).

When Urban gave his sermon at the Council of Clermont, he tapped into a religious fervor that he didn't know existed in Europe. For the rest of 1095 and all of 1096, his message spread throughout the land, raising an army larger than any ever seen before.[33] What began as an appeal by the pope to the knights of France to free the Holy Land soon turned into a major migration of knights and peasants from all over Europe, with no central leadership, traveling by land and sea to Jerusalem. Urban's call for a religious war (the word *crusade* was never actually used until the thirteenth century) served to reunite Christendom and bolster the power of the papacy.

The First Crusade was not merely a religious event, but it also represented a major sociological upheaval. It was responsible for the first organized wave of Christian violence against the Jews in Europe. While hatred of the Jews had existed in England, France, and Germany for centuries, it had been a fairly passive dislike and a tolerance based on the perception that Jews served as moneylenders and bankers for the nobility of the land. However, the peasants who flocked to the Crusade had no dealings with the Jews and considered them as evil as their Muslim enemies. Jews were considered to be less than human, or at times not even human, by these people. After all, the pope's war was aimed at infidels (or nonbelievers), and Jews in the crusaders' own countries made much more easy targets than did Arab warriors many hundreds of miles away. Plus, the Jews were vastly outnumbered, and they could not win against hordes of raging people who wanted nothing but their deaths.[34]

The day that Pope Urban set for the crusaders to begin their quest was the Feast of the Assumption in 1096. Yet many months earlier, an undisciplined brigade of enraged knights and peasants led by Peter, the Hermit of Amiens, made the journey to Jerusalem. Pope Urban had figured that several thousand knights of noblest ranking would enter Jerusalem on his behalf. He did not envision that an angry mob, a People's Crusade, would take place. But based on his sermon, the brigade consisted of approximately one hundred thousand people, many of them women and children, with few of them capable of fighting for anything other than food.[35]

Lacking all military experience and discipline, the People's Crusade marched across eastern Europe, expecting cities along the way to supply them with food and other necessities for free or at least for very low prices. The locals didn't always agree with such expectations. The People's Crusade battled the Hungarians and the Bulgarians during the journey down the Danube. Finally, still about seventy-five thousand strong, they reached Constantinople, the capital of the Byzantine Empire. Emperor Alexius, not sure what to do with this large and unruly army, ferried them all across the Bosporus into Asia Minor.

At this point, the German and French knights in the People's Crusade began arguing about leadership of their army. As usual in such disagreements, no conclusion was reached, and instead, the People's Crusade split into two groups. The Seljuk Turks took advantage of the situation and massacred the invaders.

Between November 1096 and May 1097, various armies of German, French, and Italian knights arrived at Constantinople. This new group of crusaders requested food and assistance from Emperor Alexius. But after his experience only a few months earlier with the People's Crusade, the emperor was suspicious of the new army. In return for provisions, he made the crusaders swear loyalty to him. He also insisted that they return land previously captured by the Turks. After doing so, and receiving supplies, the crusaders marched to Antioch, a major city that was between Constantinople and Jerusalem. The crusader army arrived in October 1097 and set siege to the city immediately. The city fell in May 1098, and the crusaders killed most of the inhabitants. A few days later, a Turkish army arrived, and the crusaders found themselves trapped in the same city they had been attacking for months. The crusaders defeated the Muslim army in a pitched battled outside the city a month later. Still, for months afterward, the crusaders remained in Antioch, as the leaders of various factions of the army argued over the spoils.[36]

In early 1099, the crusaders finally set out for Jerusalem. They reached the Holy City on June 7, 1099, and again put the city to siege. On July 15, the crusaders ended the siege by breaking down parts of the walls surrounding the city and entered

Jerusalem. What followed was yet another bloodbath, as the crusaders murdered nearly every inhabitant of the city, including Muslims, Jews, and even some Christians. According to accounts of the time, the slaughter was so terrible that the streets ran red with blood.

With Jerusalem and the Church of the Holy Sepulchre captured, the vows of the crusaders were fulfilled. Many of the crusaders left the city and returned home. Rumors of treasure in the city had numerous men searching for riches. Only a few minor relics were found, nothing of significance.

In *The Last Crusade*, employees of Walter Donovan found an inscribed stone tablet in 1938 in Ankara, Turkey, that told of three brothers who had participated in the First Crusades. These three crusaders had evidently located the Holy Grail somewhere in Jerusalem and traveled back with it to Turkey. Having drunk from the Grail, they were immortal as long as they stayed close to it. One hundred and fifty years passed. Finally, two of the brothers returned to Europe to inform others of their discovery, leaving the third brother to guard the chalice and its hiding place. The pair of crusaders, away from the Grail, died soon after their return to Europe, but not before telling their story to a Franciscan monk, who recorded the tale. Two tablets were inscribed with a map that showed the location of the Grail. In a race against time, Indy uses the information written down by the monk and depicted on the tablets to find the holy cup before it is discovered by the Nazis.

The Catacombs of Venice

Indiana locates an ancient church that was converted into a library. It was here that his father had disappeared while investigating the legend of the three knights. Indy discovers that there are catacombs beneath the church floor, much like the underground catacombs beneath the floor at St. Stephen's cathedral.

The term *catacombs* refers to an interconnected series of underground burial tunnels. The word was first used to describe such galleries near San Sebastiano fuori le mura, in Rome, but over the years it has come to mean any underground network used for burying the dead. Some famous examples of catacombs are the Catacombs of Rome in Italy; the Catacombs of Paris in France; the Catacombs of Alexandria, Egypt; the Catacombs of Lima; and the catacombs under St. Stephen's cathedral in Vienna.[37]

St. Stephen's was one of the oldest churches in Vienna. It was built in 1147 and enlarged over the centuries, with the work finishing in 1511. According to the legend of the three brothers, as described in the movie, two of the three returned to Europe from the Crusades 150 years after they left, which fits in perfectly with the history of the church. That one of the knights, Sir Richard, was buried in the catacombs beneath St. Stephen's made perfect sense, since catacombs were built primarily to serve as the final resting place for crusaders.

Most likely, St. Stephen's in Vienna was used as the basis for Indy's church. In the film, the catacomb tunnels are filled with water topped with oil. The presence of petroleum in the catacombs implies that the old tunnels are connected to the city sewer system. That explanation makes perfect sense when later in the scene Indiana finds a passage from the catacombs leading to a manhole in the city street. The oil floats on top of the water, since oil and water don't mix. The connection between the catacombs and the sewers also explains why the air in the catacombs is fresh after hundreds of years of supposedly being sealed from prying eyes. Exactly why Indy's torch doesn't set the petroleum fumes on fire is never revealed.

The reason for the oil in the sewers is depressingly simple. Venice has suffered from pollution for decades. Although the city has some seventy thousand residents, more than twelve million tourists visit it each year. The resulting waste and pollution are more than Venice can handle. Plus, as far back as the 1920s, petrochemical plants on the mainland have polluted the waterways of the city. The banks of many canals are covered with toxic red waste.

Agriculture runoff is also a source of pollution in the canals. Every time it rains, chemicals from farms are washed into the lagoon, creating a thick layer of algae that clogs the Venice lagoon during the summer months. This is a serious problem in dead sections of the region called *laguna morta*, which aren't cleansed by the tides of the Adriatic Sea.

Rats

Indiana and his companion, Dr. Elsa Schneider, search the underground catacombs, disturbing the large rat population that lives beneath the city.

The same catacombs are filled with rats, whose presence also makes it clear that the catacombs are connected with the city sewer system. Rats wouldn't seem to have much business being in the catacombs, though. There's no food; the bodies buried there have long since been reduced to bone.

The best-known rat species are the black rat, *Rattus rattus*, and the brown rat, *R. norvegicus*. The common term *rat* is also used in the names of other small mammals that are not true rats. Rats are distinguished from mice by their size; rats generally have bodies longer than five inches. A rat has an average life span of two to three years.[38]

In Western countries, domesticated rats are kept as pets. These are of the species *R. norvegicus*, which originated in the grasslands of China and spread to Europe and eventually, in 1775, to North and South America. Pet rats are brown rats that have descended from those bred for research. They are often called "fancy rats," but they are still the same species as the common city sewer rat. Domesticated rats tend to be both more docile than their wild ancestors and more disease prone.

The two common species of rats usually live with and near humans. Experts estimate that there is one rat for every person living in the United States.

The Black Plague is traditionally believed to have been caused by the microorganism *Yersinia pestis*, carried by the rat flea, *Xenopsylla cheopis*.[39] These rats were victims of the plague themselves.

Rats are frequently blamed for damaging food supplies and other goods. They are considered pests or vermin. They can be very destructive to crops and property. Rats can quickly overpopulate when they live in a place where they have no predators, such as in big cities. Because of this, the entire province of Alberta, Canada, has upheld and maintained a rat-free status since the early 1950s. It is even illegal to keep pet rats there.

Rats have a significant impact on food production. Estimates vary, but it is likely that about one-fifth of the world's total food output is eaten, spoiled, or destroyed by rats. Rats carry more than thirty different diseases that are dangerous to humans, including Weil's disease, typhus, salmonella, and bubonic plague.[40]

Rats are natural athletes with seemingly no end to their prowess in water, air, or land. They can swim up to a half mile and tread water for three days. While many people can swim a half mile, few of us can tread water for three straight days. Moreover, rats can jump three feet into the air, and they can plummet fifty feet down. They gnaw through anything from lead to cinder blocks.[41]

Rats also bite people. A variety of rat-control methods have been used throughout human history to either reduce or eliminate rat populations in homes, markets, farms, and industrial sites. The two most widely used methods are rat poison and rat traps, although cats and snakes have also been employed to hunt rats. Professional rat catchers can be found in many developing countries.

Because rats are nocturnal, daytime sightings of rat activity can mean that their nesting areas have been disturbed or, more likely, that there are so many of them in an area that they have run out of food to find at night and have been forced to forage in the daytime as well. As with most animals, the stronger rats push the weaker ones out into the wild to hunt. It's survival of the fittest, even among rodents.

Rats often chew electrical cables. Approximately 26 percent of all electrical cable breaks and around 18 percent of all phone cable breaks are caused by rats. About 25 percent of all fires of unknown origin are estimated to be caused by rats.[42]

The Brotherhood of the Cruciform Sword

Indy and Elsa are spotted entering the catacombs by members of a se-cret society, the Brotherhood of the Cruciform Sword. Certain that the pair are seeking the location of the Holy Grail, the Brotherhood decides to kill them by setting the catacomb oil on fire. Indy saves himself and Elsa by huddling under the overturned sarcophagus of Sir Richard. The two of them flee the tomb with the Brotherhood in close pursuit.

The Brotherhood of the Cruciform Sword is the fictional secret society dedicated to guarding the location of the Holy Grail. The phrase *cruciform sword* relates directly to and describes the swords that the knights really used. The real swords had flat bars that acted as hand guards. With the flat bars near the top of the sword, separating the base from the lethal weapon, the overall shape of a cruciform was a cross. But rather than serve as a religious icon, the crosslike sword simply protected the hand and helped the knight deflect attacks.

By definition, a secret society is an organization that conceals its activities from outsiders. The term *secret society* is often used to describe a wide range of organizations. Most people regard secret societies with suspicion and even disdain and great fear. Many secret societies, for example, have illegal objectives ranging from extortion to larceny to mass murder. In fact, secret societies having political agendas are often deemed illegal in countries around the world.

According to the Catholic Church, a religious secret society demands an oath of allegiance and secrecy. Members must be loyal to everything set forth by the society and must be loyal, in all cases, no matter what the conditions are, to one another. In addition, religious secret societies must perform religious ceremonies, including direct references to the Bible and the use of many religious symbols and rituals. Members are required to know and reference these symbols and rituals at any time and without forewarning.[43] A non-religious secret society hides its activities and rules, its uses and passwords, and the names of its members. Religious rituals and symbols, the use of the Bible, and other forms of religious worship are not mandated. But generally members must maintain utmost secrecy, often under the threat of severe punishment.

Early secret societies had arcane doctrines that were carefully concealed from anyone who was not a member. Such organizations date back to Pythagoras (582–507 BC). The Druids also formed a secret society to conceal their religious practices. As discussed earlier, the enemies of the religious order of the Knights Templar maintained that the brothers of the temple, while externally professing Christianity, were in reality pagans who veiled their pagan beliefs under orthodox terms to which an entirely different meaning was given by the initiated.[44] The most popular and widespread secret society in modern times was the Freemasons, which began in London in 1717. The most notorious secret society was the Illuminati, which has been the subject of numerous nonfiction and fiction books. The Catholic Church condemns all secret societies in no uncertain terms. Thus, members of the Brotherhood of the Cruciform Sword may have thought they were performing God's will by keeping the location of the Holy Grail hidden, but they were in direct conflict with the teachings of the church.

Venice

The Brotherhood proves to be quite dedicated to seeing Indiana and Elsa dead. They chase Indy and the beautiful professor by boat through the Venice harbor.

Venice is a city located in the middle of a large marshy lagoon approximately 208 square miles in size. Numerous deep channels run through the lagoon, making it a safe harbor for ships of all sizes. Venice proper exists on 122 small islands in the lagoon. There are approximately 150 canals in the city, and 400 bridges connecting the islands to one another. No cars are allowed in the city. You either walk or ride in a water-taxi. Gondolas are used mostly by tourists.

The harbor of Venice is deep enough to allow large tanker ships as well as American destroyers to dock fairly close to the city. Once you are away from the canals of the city, there is plenty of open water in the lagoon for a boat chase to take place; however, major cargo ships use fourteen-foot steel propellers to move through the water. The size, weight, and speed of such massive objects are enough to crush a motorboat in seconds. It's a fate that befalls several boats belonging to the mysterious Brotherhood, but Indy manages to steer his boat away from the danger.

Book Burning

After Indy's encounter with the Brotherhood ends in a truce, Indy learns from them the location of his father. The elder professor Jones is being held in Castle Brunwald, a German castle on the border of Germany and Austria. Indiana sends Marcus Brody in search of the Grail while he rushes off to find his father. It's at Castle Brunwald that Indy discovers that Elsa and Walter Donovan are working together for the Nazis. After rescuing his father, Indy travels to Berlin to retrieve Professor Jones's Grail diary. In Germany, he witnesses a book burning, a commonplace activity under Nazi rule.

On May 10, 1933, on the Franz Joseph Platz in Berlin, German students from Wilhelm Humboldt University, formerly regarded as one of the finest schools in the world, burned around twenty thousand books from the Institut für Sexualwissenschaft and Humboldt University. The books included works by Jewish authors and other so-called degenerate authors such as Sigmund Freud, Albert Einstein, Thomas Mann, Jack London, Heinrich Heine, Karl Marx, Upton Sinclair, Ernest Hemingway, Erich Maria Remarque, and H. G. Wells. As the books went up in flames, the students raised their hands in the Nazi salute. Nazi songs and anthems were sung.

Nazi propaganda minister Joseph Goebbels gave a speech to the students, stating:

> The era of extreme Jewish intellectualism is now at an end. The breakthrough of the German revolution has again cleared the way on the German path. . . . The future German man will not just be a man of books, but a man of character. It is to this end that we want to educate you. As a young person, to already have the courage to face the pitiless glare, to overcome the fear of death, and to regain respect for death—this is the task of this young generation.

And thus you do well in this midnight hour to commit to the flames the evil spirit of the past. This is a strong, great and symbolic deed—a deed which should document the following for the world to know—Here the intellectual foundation of the November [Democratic] Republic is sinking to the ground, but from this wreckage the phoenix of a new spirit will triumphantly rise.[45]

Statesmen throughout the civilized world condemned the burnings, but four years later, some of the same people stated that Hitler wasn't a real danger to freedom and democracy. Newspapers throughout the United States published editorials attacking the book burnings but rarely commented on the concentration camps springing up all over Germany. In a world of black and white, too many people looked to the Nazis and saw gray when gray did not exist.

Of all those writers who had their books burned that day, Helen Keller wrote the most eloquent response. In her letter addressed to the student body of Germany, she stated, "History has taught you nothing if you think you can kill ideas. Tyrants have tried to do that often before, and the ideas have risen up in their might and destroyed them."[46]

The book burning in Berlin witnessed by Indiana Jones in *The Last Crusade* rings false because by 1938, there were no books of any importance left in Germany to burn. Still, Indy has a book in his possession, his father's Grail diary. As Hitler approaches Indy, the ever-egotistical Führer assumes the book must be a copy of his autobiography. With a flourish, and much to Indy's surprise, Hitler autographs the book. Later, the Grail diary helps Indiana solve the mystery of the holy cup. Assuming that once the adventure is over, Indy saves the book and keeps it in good shape, today it would be worth approximately $1,000 due to the Hitler autograph.

Zeppelins

After recovering the Grail diary in Berlin, the Joneses attempt to leave Germany aboard the majestic LZ 138 zeppelin. Before the zeppelin can take off, SS colonel Vogel discovers the two men in the main cabin of the airship. Indy tosses the Nazi out the open window of the zeppelin, and he falls onto a pile of luggage. The zeppelin heads off on its route, only to be told later by a wired telegraph message that it must turn around due to "American conspirators" on board.

The LZ 138 appears to be the same type of zeppelin as that of the ill-fated LZ 129 Hindenburg that exploded in New Jersey. Unlike the Hindenburg, however, the LZ 138 has an attachment built onto the bottom of the airship frame to hold an airplane. Such attachments were commonplace among American airships of the 1930s.

Zeppelins were steel-framed dirigibles made famous by the German count Ferdinand von Zeppelin in the early twentieth century. The airship was based primarily on a flying machine designed by the Croatian David Schwarz. As was often the case with new inventions, the man who popularized the invention gained all the fame for it.[47]

Before the zeppelin came about, lighter-than-air vehicles were large balloons that served little purpose other than rising into the air to study the countryside. Zeppelin changed this by constructing a metal skeleton for his ship, made of rings and girders. Air bags were fastened to the ship's frame, giving the zeppelin more lifting power than a balloon and thus enabling it to carry more weight. This weight translated into a number of powerful engines, with propellers, which moved the airship forward. Airships with nonrigid designs were known as blimps.[48]

The earliest zeppelins were square-shaped with rounded ends. The tail assembly, used to steer the ship, was complicated. In World War I, zeppelins grew more streamline, and the tail assembly was in the shape of a cross. Each airship was lifted by a number of hydrogen-filled balloons called cells. Zeppelins used internal combustion engines fastened to the ship's gondola, which in turn was attached to the bottom of the ship's skeletal frame. The gondola carried the crew and any passengers. Steering was done by changing the direction of engine thrust and using the fins on the tail of the ship as well as elevator fins attached to the side of the outside shell.[49]

Before the start of World War I, zeppelins were used to transport people across Europe. These zeppelin flights are considered by many air historians as the first passenger and commercial airline flights. Once the war began, the German military seized control of all available zeppelins for use in bombing raids and to scout Allied positions.

The German defeat in 1918 temporarily stopped the airship business. Because Count von Zeppelin had died during the war, Hugo Eckener took over zeppelin development, and once again, flights took to the skies. While airplanes were growing increasingly popular, zeppelins were still the vehicle of choice for many wealthy Germans. In the 1930s, the LZ 127 *Graf Zeppelin* flew on a regular transatlantic schedule between Germany and North and South America.

In 1933, the Nazis gained control of the German government, and a sudden chill hit the zeppelin passenger business. The Nazis had little interest in flying people from one city to another. To them, anything that flew was a war machine. Moreover, the Nazis had seen firsthand the problems zeppelins had with fighter planes in World War I, so they were not interested in allocating resources to the lighter-than-air ships.

Still, the Nazis were experts with propaganda, and they were anxious to exploit the interest and popularity of their huge dirigibles. Eckener, no fan of the Nazi regime, refused to cooperate with the Reich's plans. Hermann Goring, the Nazi in charge of the air ministry, took over the airline in 1935. Goring called his new company the Deutsche Zeppelin-Reederei (DZR) and painted huge Nazi swastikas on the ships' fins. High in the air throughout Germany, the ships played Nazi march music and broadcast propaganda.[50]

In 1935, DZR unveiled a new zeppelin, the LZ 129. Eckener named the ship the *Hindenburg* after former German president Paul von Hindenburg. Otherwise, the zeppelin most likely would have been called the *Hitler*. The *Hindenburg* was 800 feet long and 125 feet in diameter. Compared to today's airplanes, the ship was longer than three Boeing 747s placed end to end. The gondola could carry over one hundred people and had a maximum speed of 84 miles per hour.[51]

The *Hindenburg* made its first flight in March 1936. However, the ship was inflated by hydrogen, not helium, as was previously used in dirigibles. The United States was the only country that had enough helium to inflate the multiple air bags used by zeppelins. With the Nazi party in power in Germany, the U.S. weapons embargo made it a crime to sell helium overseas. No one thought much of the switch in gases, as hydrogen fires had never been a problem on dirigibles. The LZ 129 was put on the transatlantic route, with a ticket from Germany to the United States in 1937 costing $400.[52]

On May 6, 1937, perhaps the most dramatic airship disaster up to that time took place in Lakehurst, New Jersey. The *Hindenburg* was coming in for a landing after a transatlantic flight. There were ninety-seven people onboard. Somehow, as the *Hindenburg* was landing, the tail of the airship suddenly caught fire. The hydrogen ignited, and in seconds the dirigible was engulfed in flames. Incredibly, while thirty-five people on the ship were killed in the fire, more than sixty others survived the explosion. The cause of the fire was never determined, and some people suspected sabotage.

The disaster signaled an end to the transatlantic zeppelin trade. The United States and Germany were moving closer to war, and mistrust of anything German was growing in America. On the flip side, Germans were not as anxious to board a potentially deadly vehicle to fly to the enemy zone called America. Oddly enough, despite the *Hindenburg* disaster, hundreds of people still wanted to fly overseas by airship. For the courageous, the fear of an explosion and subsequent death did not matter as much as the adventure of flying through the air to see faraway countries. But in 1940, the German government refunded the money of everyone still scheduled for a flight.[53]

When the *Hindenburg* was no more, its sister ship, the *Graf Zeppelin*, was pulled from further flights and turned instead into an airship museum. In *The Last Crusade*, the LZ airship that Indiana and his father flew on their escape from Germany existed only in the scriptwriters' minds. The last German dirigible manufactured was the LZ 130 *Graf Zeppelin* II in 1938; however, all passenger flights by dirigibles had been discontinued soon after the *Hindenburg* disaster. The *Graf Zeppelin* II was used only as a propaganda machine for the Third Reich and later as a spy aircraft over Great Britain in 1939. The airship was dismantled in 1940 for its parts and metal frames.

The Birds in the Skies

Indy and his father escape from the zeppelin in an airplane attached to the bottom frame of the giant dirigible. Unfortunately, while working the machine gun on the plane, the elder Jones manages to destroy the tail assembly. Later, he redeems himself by using his umbrella to scatter a flock of birds toward the path of an attacking airplane, causing the plane to crash. Professor Jones Sr. quotes a line from Charlemagne as an explanation for his idea. "I suddenly remembered my Charlemagne," declares the professor. "'Let my armies be the rocks and the trees and the birds in the skies.'"

Charlemagne, also known as Charles the Great, lived from 747 to 814 and was king of the Franks from 768 until his death forty-four years later. He expanded the Frankish kingdoms into an empire that included most of western and central Europe. While it is quite possible that Charlemagne originally said the line quoted by Professor Jones, there's no record of it in any reliable resource dealing with his life.

Hatay

Free from the Nazis, Indiana and his father head for Iskenderun, the city once known as Alexandretta in the center of the Turkish province of Hatay.

Iskenderun is a growing metropolis that has surpassed the provincial capital of Antakya in population. It is located on the coast of the Mediterranean Sea, on the Gulf of Iskenderun, at the foot of the Nur Mountains. The city is in the far southeast of Turkey. In *The Last Crusade*, according to the clue found at the tomb of Sir Richard, the Holy Grail is located in a canyon of the crescent moon outside of Alexandretta. All of this is well and good, except for the fact that Alexandretta was not renamed Iskenderun until 1939, a year after the adventure takes place.

The Turkish state of Hatay was a small independent state that existed for nine months between September 7, 1938, and June 29, 1939. During that short time, the state was ruled by a local sultan. Hatay was swallowed up by Turkey and became the province of Hatay on June 29, 1939.

The Phantom II

The Nazis offer the sultan of Hatay all sorts of wealth if the ruler will cooperate with their search for the Holy Grail in his kingdom. The sultan turns down all offers until he spots the car Walter Donovan is driving—a Rolls-Royce Phantom II, with a list price at the time of a little more than $20,000, which he happily takes in trade for the Grail.

The Rolls-Royce Phantom II was developed in 1929 to replace the popular Rolls-Royce Phantom I. The Phantom II made its debut at the London Olympia Motor Show in October 1929. The chassis offered some modern features. Notable was the use of half elliptic springs for both axles. In the case of the rear axle, these were underslung, which, used in conjunction with a lower frame, created a considerable reduction in the height of the car, lowering it nine inches below the Phantom I.

The engine, the clutch, and the gearbox were combined into one unit, and instead of a sub-frame being employed, the engine was bolted directly to the chassis. An innovation in the Phantom II was the use of a synchromesh gearbox for the first time. All told, 1,394 Phantom II's were built, each with a six-cylinder engine and a maximum speed of 92 miles per hour.

The Temple of the Crescent Moon

Indiana skirmishes in the Hatay desert with the Nazis, who are led by Walter Donovan and Elsa Schneider. This battle includes combat with a tank. Then Indy finds himself in the Temple of the Crescent Moon, the hiding place of the Grail. Unfortunately for Indy, the Nazis, along with Donovan and Elsa, have also made it to the Temple. Donovan shoots Indy's father in cold blood, wounding the old man badly. The only way Indy can save his father is to locate the Grail. The water from the holy goblet can heal all wounds. But to obtain the cup, Indy must traverse three traps set by the crusaders hundreds of years earlier to kill anyone seeking the Grail for personal gain. The first trap has already killed several native guides who were helping the Nazis. The men were beheaded while passing through a narrow corridor. Indy's only clue is a line from his father's Grail diary: "The breath of God—only a penitent man will pass." Indy realizes that a penitent man is a humble man, and that such a man would bow in the presence of God. Just in time, Indy ducks and rolls forward, barely missing being decapitated by a deadly triple pendulum.

The second clue involves a religious man knowing the name of God. A maze of lettered stepping-stones lies in Indy's path. The wrong step on the incorrect letter will send Indy plunging into the far depths of the temple. Only just in time does Indy remember that God's name in Latin begins with an I, not a J.

The third test requires Indy to make a leap of faith. This test relies on a bridge that is visible only when viewed from a variety of different positions but not when one is looking straight down at it, creating a perfect optical illusion.

The impressive Temple of the Crescent Moon is real, but it isn't located in Turkey. Instead, the building is located in the ancient ruined city of Petra, Jordan. The only approach to this particular temple is through a dark, narrow gorge known as the Siq—the Shaft—a deep split in the sandstone rocks of the region. At the end of the Siq stands Petra's most spectacular ruin, Al Khazneh, the Treasury, cut right out of the sandstone cliff. Recently, Petra was named one of the New Seven Wonders of the World.

A pendulum is a weight attached to a pivot point by a string or a rope so that it can swing freely. When the pendulum is displaced from its place of rest, the restoring force will cause the pendulum to oscillate around the equilibrium position. A gravity pendulum is a weight connected to one end of a string (with no mass). The other end of the string is connected to the ceiling. When given a push, the weight will swing back and forth under the influence of gravity over its lowest point.

Galileo Galilei was not the first man to notice the properties of a pendulum, but his studies of the pendulum influenced the scientists and clockmakers who followed him. Galileo discovered that the period of a pendulum was independent of the mass of the weight or the amplitude of the swing. He also deduced that there was a direct relationship between the square of the period and the length of the string. Galileo's work with pendulums led to the invention of the first pendulum clock by Christiaan Huygens in 1656. While the deadly pendulum makes exciting movie action, it seems unlikely that the three brother crusaders could have built such a trap in the eleventh century.

As for the second trap Indy encounters, how did he know to use the letter *I* and not *J*? The letter *J* was unknown in any alphabet until the fourteenth century. It was added to the Latin alphabet to differentiate from the letter *I*, which up to that time not only was used as a vowel but also had the consonant value of *Y*. In Latin, either symbol had the sound of *Y*, so the Latin pronunciation of both *Ianuarius* and *Januarius* was "Yanuarius."

Thus, when Indy begins to spell the name of God, he has forgotten that for the crusaders, the letter *J* didn't exist. Thus, *Jehovah* would have been spelled with the letter *I*. But since the letter *J* didn't exist during the time of the crusaders, how is it one of the letters on the maze on the floor in the Church of the Crescent Moon?

Plus, God's name wasn't spelled "Jehovah" using a *J* or an *I* until it was used in the King James Bible in 1611. That's hundreds of years after the crusaders found the Holy Grail and placed it in the Church of the Crescent Moon. In the twelfth century, God's name was symbolized by the Tetragrammaton, YHWH, a series of four Hebrew consonants. The English transliteration of those letters is yodh, he, waw, and he. *Tetragrammaton* is Greek for "the four letters." The name of God was pronounced freely during the Age of the Prophets, but it is no longer pronounced other than in very important religious moments. According to Jewish custom, there are strict rules for writing God's name. Once God's name has been written, it can neither be erased nor discarded. It must always be stored.

Various translations have been given for the Tetragrammaton, including "the Lord" and "the Name." The correct pronunciation has been lost in the thousands of years since it was last properly said aloud. Many occult scholars feel that the correct pronunciation of God's name is the key to understanding the meaning of the universe. This notion was the basis for the movie *Pi* and has been used in numerous occult novels, including *The Black Lodge* by Bob Weinberg, one of this book's authors. But no possible pronunciation of the Tetragrammaton would have produced the much later word *Jehovah*, in either spelling.

As for the third test involving the rock bridge, we have to assume that this bridge existed when the crusaders first came upon the Temple of the Crescent Moon, because building it across the deep gorge would have been nearly impossible. Building it to blend in with the rock face on the far side of the narrow canyon would have been a long, laborious chore. But what else did the guardian of the Grail have to do with his time? No doubt, he spent hundreds of hours making sure the optical illusion was perfect. And it was, requiring Indiana to step out into God's hands to save his father's life.

The Jerusalem Cross and the Third Brother

Indiana manages to figure out the meaning of the clues, which makes it possible for him to enter the Grail room. Here waits the third brother, wearing a Jerusalem Cross on his breastplate and having stood patiently for nine hundred years, with the true Grail on a nearby ledge.

The knight whom Indiana Jones encounters is still wearing his knightly robes, which says something about the quality of laundry service available in the caves. After all this time, his clothing should have turned to dust. Since no one else is ever seen in this odd Temple of the Crescent Moon, we are left to speculate that the knight lives by himself in the ancient caves and that the Grail has given him immortality. His service takes religious dedication to an entirely new level. But perhaps that is the true meaning of the Holy Grail.

Any man wanting to become a crusader had to undergo a specific ceremony. This ritual changed over the course of a hundred years, but the basics remained the same throughout Medieval times. The potential crusader sought out a leader of the church, usually a priest, but for the nobility, a bishop or a cardinal. The soldier swore that he would carry out an armed pilgrimage to the Holy Land and free holy places. After he concluded his pledge, the man was given a cloth cross to wear that signified he was a crusader.[54]

The crusader's cross represents Christ's command to spread the Gospel around the world, a mission that was to begin in Jerusalem. It was later adopted as part of the coat of arms of the short-lived Kingdom of Jerusalem established by the crusaders, which lasted from 1099 until 1291. The crusaders cross consists of five crosses. The most common interpretation of the five crosses is that they represent the five wounds of Jesus on the cross—small crosses for the hands and the feet, and the large central cross for the spear wound in his side.

The Crusades were a deadly business, yet hundreds of thousands of people participated in them. By fighting in the Holy Wars, crusaders received an indulgence, which was a guarantee by the church that if a man was truly sorry for his earthly sins, as demon-

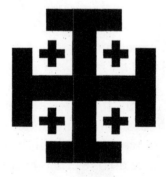

The Jerusalem cross

strated by going to war against pagans, he would be granted forgiveness when he died. Even the richest nobles were deeply religious, and an indulgence was like a "get into heaven free" card. It
was an offer that most people could not refuse.[55]

In the Middle Ages, when men became monks, they made
three vows. The first was the vow of poverty. This meant that the
person had to give up all personal possessions and forsake all
wealth. The second vow was to remain chaste and pure. This became known as the vow of chastity. The third vow was a commit-
ment to obey the leaders of the monastery and the church. This
became known as the vow of obedience. Many crusaders who witnessed the massacres, the tortures, and the multiple atrocities
committed during the Crusades became monks when they returned to England. Perhaps the best-known book series about a
crusader who became a monk is the Brother Cadfael series by
Ellis Peters.

Certain extremely religious or extremely disillusioned crusaders
led religious lives without belonging to a monastery or a church.
These men lived in small dwellings or even in caves, devoting their
lives to simple prayers and sometimes to healing the sick. These
men were known as hermits. During the Middle Ages, hermits
were thought to be holy and wise, although some hermits were
neither.

The third brother found by Indiana Jones in the Temple of the
Crescent Moon in Turkey was most likely a knight who had, in his
own way, become a hermit. He had forsaken normal life to remain
and guard the Holy Grail until someone worthy enough came to
take it from him.

The Cups of Kings and Carpenters

*When Walter Donovan and Dr. Schneider follow Indiana into the
knight's chamber, they find themselves confronted by dozens of cups. The
question is, which cup is the true Grail? The knight tells them that while
drinking from the true Grail will bring them everlasting life, a drink
from the false Grail will take it from them. Dr. Schneider selects a gold*

cup covered with jewels as the true Grail. The appearance of such a fabu-lous cup appeals to Donovan's somewhat twisted view of religion. To him, Jesus is king of kings and the lord of hosts—not a humble man, but one of immense power. To Donovan and Schneider, power means wealth. Thus, Donovan takes the golden cup and drinks from it. But he has made the wrong choice. In an instant, he ages and turns to dust. Indiana realizes that Jesus, the champion of the poor, would never drink from a vessel cov-ered with jewels.

Indiana selects the true Grail, a plain wooden cup, the cup of a carpen-ter who lived and worked with the poor. He drinks from the cup without harm, then fills the Grail with water and brings it out to his father to drink. One gulp of the water and a splash of the holy water on the bullet wound heals Dr. Henry Jones.

Why wasn't a drink from the cup enough to heal the elder Jones? Most likely, Indy was merely being careful. According to many of the Grail legends, the knight Sir Gawain was healed from a wound caused by a magical lance by pouring water from the Grail on it. If a drink from the Grail will stop a man from dying, splashing some water from that same cup on a deadly wound seems like good insur-ance. Better to be safe than sorry.

The ancient knight who guards the Grail might be a holy man, but that does not make him a merciful one, especially when it comes to those who plan to use the power of the Grail for evil means. Nor is the God who killed the villains who opened the Ark of the Covenant in *Raiders of the Lost Ark* a forgiving God.

We can't help but wonder about the claim that Jesus' cup from the Last Supper was a plain, unadorned chalice without markings and devoid of any gold or silver decorations. First and foremost, we need to establish what sort of meal the Last Supper was. Was it merely a gathering of Jesus and his disciples, or was it a more im-portant meal, as believed by many, the Seder meal held on the first night of the Jewish holiday of Passover? Based on evidence found in the Bible, it seems like the second choice is the right one.

In the New Testament, the first four books are known as the gospels. The first three gospels, those of Matthew, Mark, and Luke, are so similar that they are often called the *synoptic* gospels. These gospels tell many of the same stories about Jesus, follow pretty much the same order, and even use the same words. The synoptic gospels make it clear that the Last Supper was a Seder dinner, most likely the first night of Passover. The Gospel according to John treats the Last Supper as a meal the night before Passover. Many Christians, including many Catholics, as well as many Evangelicals, believe that the Last Supper was a Seder.

Now, the cup normally used for the various rituals at a Seder is usually the finest cup owned by the lead participant in the ritual dinner. The Seder is meant to be a celebration of the Jews leaving Egypt. It's a sacred and very important holiday on the Jewish calendar, and it has been for thirty-five hundred years. Passover may be the oldest continuous religious festival in the world. One of the most important lines in the Seder is recited by the lead participant, "I do this because of what the Lord did for me when I came out of Egypt" (Exodus 13:8). The words are spoken in present tense, because Passover is a holiday that celebrates the continued importance of God in Jewish life.

As it is now, so it was two thousand years ago. It seems extremely unlikely that the cup Jesus used for such an important dinner would be so plain. According to Joe Zias, a former curator of the Israel Antiquities Authority, the Last Supper was unquestionably a Jewish Passover Seder in which the finest tableware available would have been utilized. When asked about the possibility of a clay cup being used by Jesus for the Last Supper, as shown in *The Last Crusade*, Zias stated, "It's such a pedestrian piece of pottery that you definitely wouldn't use it for any type of ceremonial function."[56]

Of course, with all of this taking place two thousand years ago, not everyone agrees. Dr. Stephen Pfann, an American Bible expert, believes that Jesus did use a simple clay cup. Pfann bases his conclusion on research done on the pottery of the Essenes, a sect of extremely devout Jews who lived around the same time as Jesus. Pfann thinks the Essenes used a communal clay cup for all of their ritual meals due to their extreme poverty. Pfann thus agrees that the plain, undecorated cup used by Indiana Jones in *The Last Crusade* was the correct one. But no one knows for sure.[57]

The Great Seal

Despite being warned by the guardian of the Grail not to take the Holy Grail past the Great Seal on the floor of the temple, Elsa grabs the relic from Indiana's father and does exactly that.

Elsa is an extremely incompetent Nazi, a trait displayed earlier when she expresses second thoughts about double-crossing Indy in Berlin. She is a rarity, an evil villain with the trace of a conscience.

Exactly what the Great Seal is and why the Grail cannot cross it are never made clear. There is no mention of a Great Seal in any of the major accounts of the Holy Grail, and the Great Seal only vaguely resembles any occult symbol. It is pure Hollywood, a mystery that exists only to ensure that Indy does not return with the Grail to the mundane world.

The seal consists of two concentric circles with lettering between them, with a "Grail cross" at the center. The cross is the emblem of the Brotherhood of the Cruciform Sword, whose members Indy encountered earlier in this adventure. They serve the guardians of the Grail, which ties them into the Great Seal, the ultimate guardian of the cup. The design of the Grail cross is a Latin cross

The Grail cross

with a "chalice" shape over it. Exactly what the words are between the two circles is less clear, although one word appears to be "omnipotent," written in English.

The Grail Cross

When Elsa crosses the Great Seal, the Temple begins to shake as if hit by an earthquake. The Nazis' Arab mercenaries flee. The remaining Nazis are killed. The Grail falls into a newly formed crevice. Elsa desperately tries to snatch it. Despite Indy's attempt to save her, she refuses to give up the Grail and falls into the abyss. Indy almost joins Elsa in the unknown but is saved by his father, who tells him, "Indiana. . . . Indiana. . . . let it go."

Indy, Henry Sr., Indy's friend Sallah, and Marcus Brody ride off into the sunset. It's revealed here that Indiana's real name is Henry Jones Jr. and that he took the nickname Indiana from the family dog.

The notion of a Seal keeping the Grail from being taken from the Temple of the Crescent Moon is a twist on the notion of the Seal of King Solomon. In various legends that were popular during the Middle Ages, Solomon's Seal was engraved on a large ring worn by King Solomon, and it gave him the power to control demons. Those powers included trapping these demons (or djinns, in Arabic legends) in bottles. At the center of the ring was the Star of David surrounded by two circles. Inside the inner and the outer circle, all types of magical signs and sigils were engraved. While it's never specifically stated, the power of the ring came from the center symbol, an expression of God's name.

PART 4

YOUNG INDIANA JONES

Create a movie icon and raise a thousand questions; that was what happened to George Lucas after producing the first two Indiana Jones movies. Although the Jones movies were based on the old 1930s and 1940s serials, exactly how did one become an adventuring archaeologist searching the world for lost historical treasures? Who were the people who influenced Indiana Jones as he was growing up? It was with these questions in mind that producer Lucas and director Steven Spielberg approached the third Indiana Jones movie. The two men decided that the beginning of the film should reveal something of Indy's background, from how he got the scar on his chin to the origin of his famous fedora.

For the role of young Indiana, Harrison Ford suggested River Phoenix, who had played his son in the film *The Mosquito Coast*. Ford felt that Phoenix looked the most like he did as a teen. While filming the adventures of young Indiana Jones in *The Last Crusade*, Lucas was struck by the idea that such a character would be perfect for a TV series. For years, he had been courted by the major networks for a spin-off series from one of his successful movie franchises. *The Young Indiana Jones Chronicles* seemed like a perfect fit.

The series featured an elderly Indiana Jones recounting adventures from his childhood to interested bystanders. The role of Indy at age ninety-three was played by George Hall. The part of Indy as a young boy was filled by Corey Carrier. River Phoenix turned down the title role of the TV series. He died tragically from a drug overdose in 1993. Instead, Sean Patrick Flanery played Indiana as a teenager.

The adventure sequence featuring a young Indiana Jones that opened *Indiana Jones and the Last Crusade* was set in 1912, when Indy was thirteen years old. The *Young Indiana Jones* TV series took place in the time slightly before and slightly after that date. The early 1900s was a remarkable period in history that saw the beginning of the modern era, with the incredible growth of the Industrial Revolution, the horrors of world war, amazing advances in science, and the rise of political turmoil across the globe. It was one of the most exciting periods in history, and in his series Lucas wanted to take full advantage of the many distinguished people who were alive during these years.

Lucas fashioned *The Young Indiana Jones Chronicles* not only as adventures but also as slices of history. He shot the series on location around the world. Episodes focused on Indy at ages ten and sixteen. In one episode, a fifty-year-old Indy, as portrayed by Harrison Ford, told the story. Over time, viewers learned that as Indy grew older, he not only had a daughter, but he also had a grandson.

All of the episodes featured Indy meeting and working with celebrities of the early twentieth century. For example, in the first adventure Indy meets Lawrence of Arabia years before T. E. Lawrence earned that nickname. In other stories, Indy encounters Pablo Picasso, Henri Rousseau, Teddy Roosevelt, Albert Schweitzer, Princess Sophie, Eliott Ness, Al Capone, Mata Hari, and Ernest Hemingway. In one adventure Indy goes on safari with Teddy Roosevelt; in another, he is saved from the threat of Al Capone by a young Eliott Ness.

The problem with the series was that Indy encountered *too* many famous people. In the course of forty-four adventures, he ran into seemingly every important person who was alive from 1910 through 1920. At times, it seemed that everyone Indiana met had stepped right out of a history textbook. And his encounters always revealed what made the celebrities notable.

Sigmund Freud gave him psychological advice. Albert Schweitzer nursed him back to health in Africa. Howard Carter, the archaeologist, showed him how to excavate mummies. It was all very informative and educational, but too often it seemed staged, unbelievable, and, most of all, dull. A series aimed at children and teenagers, it suffered from too lofty a vision and not enough action.

Another aspect of the series that alienated audiences was the portrayal of Indiana as an old man. Many of the early episodes of the series were ruined by the sometimes preachy remarks made by the ninety-three-year-old retired professor. Too often, the shows began with old Indy talking to complete strangers, an aspect of old age that most people did not find amusing. Plus, while the series was presented as being historically accurate, sometimes facts of history were bent out of shape to enable Indy to meet famous people in locations where they might not have ever been or who in actuality were much older than presented in the stories.

The series ran for only six weeks in the spring of 1992, at which time ABC dropped it from its schedule, leaving twelve episodes not shown. It was, however, typical with Hollywood, where big names meant a lot more than anything else, that the show was nominated for eight Emmy Awards. It won five. Lucas persuaded ABC to try the series for a second season in the fall of 1992.

The second season was a mix of episodes from the previous season that had not aired, along with new stories. The introductions featuring old Indy got shorter and shorter and then were dropped entirely. In an attempt to capture the Indiana Jones movie audience, the first episode of the second season featured a brief appearance by Harrison Ford, playing Indiana at fifty. Still, ratings were terrible, and after twenty-one adventures, the series was canceled. The second season, the series won another six Emmys. The series was revived for a third and final time in 1996 on the USA network, where several never-before-shown episodes were featured along with four made-for-TV movies.

The entire series was promised as a video set, but only twelve tapes were released, leaving a number of adventures unavailable. DVDs of all the episodes, including numerous documentaries on the historical figures featured in the show, were first released in 2007 and will continue in 2008, to tie in with the release of the fourth Indiana Jones feature film.

A number of important filmmakers either wrote or directed episodes of the series, including Frank Darabont, Terry Jones, Simon Wincer, Carrie Fisher, and Vic Armstrong. George Lucas was often given credit as "story by" in a number of episodes because he served as the creative consultant for the entire series.

Was the show worth watching? Was it historically accurate and, at the same time, entertaining? Was it really possible for a young Indiana Jones to meet so many important historical figures during his travels? Were the shows fact or fantasy? We'll try to answer some of those questions, and others, in the rest of this part.

Indy's First Adventure: Lawrence of Arabia and the Curse of King Tut

In his first recorded adventure, young Henry "Indiana" Jones Jr. departs the United States in 1908, when he is nine years old, traveling with his father, Henry; his mother, Anna; and his tutor, Miss Helen Seymour, on a worldwide lecture tour that begins in Egypt. Indy and Miss Seymour travel up the Nile, and in the Valley of the Kings they meet archaeologist T. E. Lawrence, who would become well known years later as Lawrence of Arabia. Indiana and his tutor are also introduced to archaeologist Howard Carter, who will gain fame when he discovers the tomb of King Tutankhamen in 1922. While trying to solve a murder mystery, Indy is kidnapped and almost sold as a slave at a market in Marrakesh. Using his wits, young Indiana manages to escape and is reunited with his parents.

It's appropriate that young Indiana's first adventure takes place in Egypt's Valley of the Kings, one of the most mysterious and fascinating locations in the world. It's also the spot where many cultural historians believe modern archaeology was born. Exactly what made this location so unique and what brought together three such mismatched people as young Indiana Jones, Howard Carter, and Lawrence of Arabia?

The region known as the Valley of the Kings is located on the west bank of the Nile River, approximately five hundred miles south of the Mediterranean Sea. The valley is part of a much larger ancient metropolis that many historians also label the Valley of the Kings. On the east bank of the Nile River, directly across from the valley, is the modern city of Luxor. More than three thousand years ago, the capital city of ancient Egypt, Thebes, covered both banks of the Nile, where Luxor and the Valley of the Kings now stand. On the east bank was the city of the living, the dwelling place of the pharaohs. On the west bank was the city of the dead, the Necropolis of Thebes, where the kings and the most powerful nobles of the kingdom went to their eternal reward. Thirty centuries ago, Thebes was a city of the living and the dead.

Over a period of five hundred years, from the sixteenth to the eleventh century BC, the Necropolis of Thebes served as the burial ground for the rulers of the New Kingdom of Egypt. The city of the dead stretched over two entire small valleys on the west bank. The East Valley was where most of the tombs of the kings were located, and the West Valley was the burial area of powerful nobles and priests. There are thousands of tombs located in the Necropolis of Thebes. Archaeologists have explored and numbered approximately eight hundred of them.

When Indy and Miss Seymour visited the Valley of the Kings, many of the ancient tombs hadn't been opened, much less explored. Modern archaeology was a fairly new science, a little older than a quarter century, and thieves still haunted the dark corridors of the Necropolis. Fortunes could be made stealing from the dead, as had been the case for more than two thousand years.

Let's look at some of the people credited with inventing archaeology. People have always been interested in the past. The excavation of ruins and monuments has taken place for thousands of years. Herodotus, the great Greek historian, visited the pyramids in 450 BC and wrote about them in his *Histories*. Yet it wasn't until Flavio Biondo, an Italian historian in the Renaissance who was responsible for coining the term *the Middle Ages*, that anyone tried to come up with a method for studying and cataloging ancient ruins.

In 1430, Biondo and several other historians started to explore the deserted and overgrown ruins of ancient Rome. The Eternal City was little more than an empty field. Biondo spent years documenting the geography and the topography of the buried metropolis. His books *Rome Restored* and *Rome Triumphant* revived interest in the ancient city and described the glories of the fallen center of the Roman Empire. Biondo is first on our list. For his work describing ancient Rome, Biondo must be credited as one of the founders of archaeology.

Still, it wasn't until the nineteenth century that people began to spend time systematically studying the past based on existing physical remains. The English were among the first to recognize that archaeology was a science and required scientific methods. The first people involved in such research were clergymen who took careful notes about their parishes. Such men noted the locations of ancient monuments and barrows throughout the countryside that might otherwise have been lost.

Several wealthy Englishmen began to investigate the history of the countryside they inherited from their ancestors. Second on our list of the men who were responsible for the modern science of archaeology is Richard Colt Hoare. In 1812, he published a book titled *Ancient Historie of Wiltshire*, which recorded his explorations of the prehistoric barrows in the region. Hoare is credited as being the first researcher to use a trowel for careful excavation.

Third on our list is Napoleon, who invaded Egypt in 1798 in hopes of making it a French colony. Napoleon took with him 500 civilians, including 150 scientists. They were entrusted with conducting a detailed study of ancient Egypt. Their collected work was published as *Description de l' Egypte* (1809–1822). Featuring numerous drawings and paintings of the country and its ruins, the work stimulated a revival of interest in ancient Egypt, especially in England. Equally important, during the French Army's stay in Egypt, the Rosetta stone was found in a harbor on the Egyptian coast of the Mediterranean Sea.

The text on the Rosetta stone was a decree from King Ptolomy IV concerning taxation, dating from approximately 200 BC. What made the stone important was that the decree was written in ancient Egyptian (Coptic), hieroglyphics, and ancient Greek. Since scholars could read and understand both Coptic and ancient Greek, they were able to use the Rosetta stone to decode the previously undecipherable hieroglyphics. Still, it took French linguist Jean-François Champollion twenty-five years to fully unravel the meaning of the ancient text. Deciphering the Rosetta stone proved to be the key to the study of ancient Egypt, and Champollion was considered by many to be the father of Egyptology.

Napoleon's attempted conquest of Egypt failed, and a number of Muslim rulers took charge of the kingdom for the next eighty years. During that time, many British lords visited the ancient land and opened many of the ancient tombs, searching for relics to bring back to England. In certain cases, they hired agents to do the work for them. Perhaps the most illustrious of such men was Giovanni Battista Belzoni, whom we discussed in detail earlier in this book.

In 1882, the British invaded Egypt to put down a nationalist revolution that was hostile to foreign powers. They took control of the country to prevent any other European nation from gaining control of North Africa, and they remained there until 1922. During this time and shortly afterward, numerous British scholars traveled to Egypt to study the tombs of the pharaohs. These were the years that archaeology was transformed from a wealthy gentleman's hobby into a true science.

The next, and perhaps the most important, man our list is William Flinders Petrie. Petrie, nicknamed "the father of modern archaeology," was a self-taught Englishman who began to excavate the Great Pyramid at Giza in 1880. A careful and meticulous worker, he studied every bucket of sand and soil, looking for clues on how the pyramid had been constructed. Petrie fully developed the concept of seriation in archaeology, which permitted accurate dating of relics long before carbon dating was invented. It was Petrie who introduced the scientific method to archaeology. Over the next forty years, he worked on sites throughout Egypt and Palestine, making numerous discoveries about the ancient civilizations that had flourished in the Middle East thousands of years earlier. Both T. E. Lawrence and Howard Carter studied with Flinders Petrie in Egypt.

Who were Lawrence and Carter? They both were English, wanted to study ancient Egyptian life, had worked with Flinders Petrie, and left their mark in the public eye years after encountering young Indiana Jones. How did they, each in his own way, influence Indy's life?

Lieutenant-Colonel Thomas Edward Lawrence, CB, DSO, was born on August 16, 1888. He became world famous as Lawrence of Arabia. As a British soldier, he acted as a liaison between the British Army and the Arab leaders who revolted against German rule during the years 1916 to 1918. Lawrence's public image was greatly promoted by the work of journalist Lowell Thomas, who sensationalized the story of a lone British soldier among a horde of bloodthirsty Arabs. Lawrence helped to promote his own self-image in his autobiographical book *The Seven Pillars of Wisdom*.

Lawrence was born in North Wales, the son of a baron. In 1909, he went alone on a three-month walking tour of the crusader castles in Syria. He walked more than a thousand miles on foot. Evidently, this was when he met young Indiana Jones.

Lawrence graduated with first-class honors from Oxford in 1910. He was offered a job serving as an archaeologist in the Middle East and soon was working on excavations in northern Syria. For a short time he worked with the eminent Flinders Petrie in Egypt.

In January 1914, Lawrence and archaeologist Leonard Wooley were recruited by the British military to update a map of the Negev desert for the army, with special attention given to water sources. Once World War I broke out, Lawrence went to work for British Intelligence in Cairo. In October 1916, he was sent out into the desert to report on various Arab nationalist movements.

Lawrence fought with Arab troops in guerrilla action against the Ottoman Empire. His principal success in the war was convincing the Arabs to help the British. Adopting many Arab customs and traditions, Lawrence became a confidant of Prince Faisal, one of the most important Arab leaders.

Lawrence worked for a short time with war correspondent Lowell Thomas, who with his cameraman shot a lot of footage of Lawrence riding in the desert. Thomas used the film to make a movie about the war, and he toured the world with it. The movie made Lawrence extremely famous. After the war, Lawrence was made a Companion in the Order of the Bath and was awarded the Distinguished Service Order. In 1935, Lawrence died from injuries suffered in a motorcycle accident when he was forty-six.

Peter O'Toole, who played Lawrence in the epic movie *Lawrence of Arabia*, was six feet, three inches tall. In real life, Lawrence was only five feet, five inches.

The other archaeologist Indiana met on his trip to the Valley of the Kings was Howard Carter. Carter was born in May 1874 and was an English archaeologist who became a household word due to the most intriguing curse of the twentieth century.

Carter, the youngest of eight children, received no formal education, but his father, an artist, taught him the basics of drawing and painting. In 1891, when Carter was only seventeen, he began his career as an archaeologist copying Egyptian paintings and writing, and later he studied with William Flinders Petrie.

At age twenty-five, Carter went to work for the Department of Antiquities of the Egyptian Ministry of Culture, the agency responsible for the conservation, protection, and regulation of all antiquities and archaeological excavations in Egypt. It was during this time that he erected the first electric lights in the Valley of the Kings in various tombs and at the temples at Abu Simbel.

Several years later, Carter met George Edward Stanhope Molyneux Herbert, the Fifth Earl of Carnarvon, an extremely wealthy British aristocrat and amateur Egyptologist. The Earl, impressed with Carter's work, put the archaeologist in charge of all future excavations. It was during this period that Carter supposedly met a young Indiana Jones. Carter discovered six tombs in the Valley of the Kings while working for Lord Carnarvon. But that wasn't enough. Carter had become obsessed with finding the lost tomb of an obscure pharaoh named Tutankhamen.

Carter searched for the tomb for more than fifteen years. The luck struck in 1922, when finally he unearthed the steps leading to Tutankhamen's tomb. Carved on the entrance of the tomb were the words "Death Shall Come on Swift Wings to Him Who Disturbs the Peace of the King." On November 26, with a number of dignitaries on hand, Carter opened the best-preserved and most intact tomb ever found of an Egyptian king. The tomb contained an amazing collection of treasures, including a stone sarcophagus with three gold coffins nested within one another. Inside the final one was the mummy of the boy-king Pharaoh Tutankhamen. It took more than ten years to catalog all of the antiquities found in the tomb.

A few months after the discovery of the tomb, stories began to circulate about the Curse of King Tut. The curse became a national sensation when Lord Carnarvon died unexpectedly in Cairo. By 1929, eleven people connected with the discovery of the tomb, including Carter's personal secretary, had died of unnatural causes. By

1935, the curse allegedly had claimed twenty-one victims. Yet no one could explain how Howard Carter remained alive and in good health.

After finding King Tut's tomb, Carter retired from archaeology field work and became a private collector of Egyptian antiquities. He toured the United States in 1924 and gave a series of lectures in New York City that were extremely well attended by enthusiastic crowds. Carter died on March 2, 1939, at the age of sixty-four.

When Indy was taken prisoner and given to white slavers to be sold in the slave market of Marrakesh, fact and fancy combined. The slave market in Marrakesh was not stopped by the government until 1912. It was quite likely that a young white American boy would have fetched a good price from the slavemasters who haunted the market. And, once sold, Indy would have never been seen again.

When it comes to the business of trading human lives for money, the East African Arab slave trade is probably one of the oldest. It began long before the Europeans gathered slaves for sale in the New World. Female slaves generally ended up in the Middle East as concubines; the male slaves had the unfortunate, to put it mildly, experience of becoming eunuchs before spending the rest of their lives doing manual labor and fighting in wars. In the period stretching from AD 650 until approximately AD 1900, the Arab slave trade thrived in North Africa. Reference works confirm slave owning in Arabia and Yemen through 1920. Laws prohibiting slavery were established numerous times in Mauritania—even as recently as 1981—but without criminal penalties associated with the laws, slave traders have continued to sell human lives.

Passion for Life: A Party in France, a Safari in Africa, a Romance in Vienna

After his adventures in Egypt, nine-year-old Indy travels with his parents to Paris. Visiting the Louvre, he meets a fifteen-year-old artist named Norman Rockwell. Promising to show Indy the place where real artists live, Rockwell takes him to the Montmartre section of Paris. There, in a café, they meet another young artist, Pablo Picasso, who is arguing with a much older Edgar Degas about who is a better painter. Afterward, Indy and Rockwell accompany Picasso to a dinner party that Picasso is throwing for elderly artist Henri Rousseau. On the way to Picasso's studio, Indy learns about the new style of art known as cubism.

Who were these great artists and did the party that Indy and Norman Rockwell attend really take place? Or was it merely a gimmick that was used to make an interesting hour of television?

The Louvre Museum in Paris, France, is one of the oldest, largest, and most impressive art galleries and museums in the world. In 2006, the Louvre was visited by more than 8.3 million people, making it the most popular art museum in the world.

Since its beginnings in the late twelfth century, the Louvre has dominated central Paris. In 1190, an earthen rampart was built around Paris, Europe's biggest city, to protect it from Viking raiders. King Philippe Auguste, wanting to reinforce his defenses, built a fortress on the western edge of the city that became known as the Louvre.

In 1535, King François I had the original building torn down and a new palace built in its place. This palace also became known as the Louvre. Work on the building was finished by King Henri II. Over the next several centuries, a succession of kings expanded the palace, adding numerous pavilions and wings to the building. Following the completion of Versailles, however, the French kings lost interest in the Louvre. With the French Revolution, the Louvre was transformed into a museum for the people of Paris. It has remained such ever since. The Louvre holds several of the world's most prestigious works of art, including Leonardo da Vinci's *Mona Lisa*, the *Winged Victory of Samothrace* depicting the goddess Nike, and Alexandros of Antioch's *Venus de Milo*.

That Indy visited the Louvre on his first trip to Paris is no surprise. For anyone interested in ancient art and sculpture, it is a must-see. In a time before television, radio, and all but the most primitive motion pictures, the Louvre was one of the few places Indiana could visit for hours and be constantly entertained by relics from the past.

Montmartre, where Indy and Rockwell meet Picasso, is a hill 425 feet high in the north of Paris, a part of the Right Bank that is primarily known for the white-domed Basilica of the Sacré Cœur on its summit and for being an artist district. The neighborhood is also famous for its many nightclubs, including the notorious Moulin Rouge and its rival, Le Chat Noir.

Montmartre has a rich and varied history. The site is one of the oldest known settlements in Europe and one long invested with religious significance. Because it overlooks the region, druids likely considered it sacred. In AD 250, St. Denis, the bishop of Paris and later the patron saint of France, was beheaded on the mountain, leading people to call it the "Mountain of the Martyr," or Montmarte.[1]

It was Napoleon III and his city planner Baron Haussmann who accidentally created an artist district in Montmartre. The two men wanted to make Paris the most beautiful city in the world. When Haussmann redesigned the center of the city, he gave the best sections of land to his friend. This drove the poor people to areas not part of the city proper, including Montmartre.

Montmartre was not subject to Paris taxes because it was located beyond the city limits. The wine made by local nuns attracted merrymakers of all kinds, and by the twentieth century the region became famous for licentious entertainment, including the Moulin Rouge and Le Chat Noir.[2]

The Church of the Sacre Coeur was built on Montmartre from 1876 to 1912, with donated public money. Its white dome was a distinctive, very visible landmark in the city, and its courtyard became famous as a place where artists set up their easels each day to paint. They still do.

In the late 1800s, numerous artists including Johan Jongkind, Camille Pissarro, Pierre Brissaud, and Vincent Van Gogh moved to Montmartre. By the beginning of the twentieth century, Montmartre and its rival neighbourhood on the Left Bank, Montparnasse, had become the two main art centers of Paris. In 1904, Pablo Picasso moved into a commune on the hill. He remained there until 1909, when he could afford better quarters.

The colony attracted the giants of the art world, from Matisse, Renoir, Degas, and Toulouse-Lautrec to Duchamp-Villon, Deraine, Valadon, Utrillo, and Steinlen. It also led to the formation of influential associations such as Les Nabis and the Incoherents.[3]

Indiana's companion on this trip to Montmartre was Norman Percevel Rockwell, a twentieth-century painter born on February 3, 1894. Unfortunately, there's no record of Rockwell ever being in Paris before World War I. He did visit Paris in 1923 and studied art there for a short time. Although Rockwell greatly admired Picasso's work, there's no record of the two artists ever meeting.

Rockwell was born in New York City, the son of Jarvis Waring and Ann Mary Rockwell. He was a gifted student. When he was sixteen, Norman left high school and attended the New York School of Art. Later, he went to the National Academy of Design and the Art Students League. *St. Nicholas* magazine and the Boy Scouts of America's magazine, *Boys' Life*, were among his first clients, and some of his first paintings were done for those magazines. Rockwell claimed that the composition in each of his works was organized to form a triangle in some way.

At age nineteen, Rockwell became the art editor for *Boys' Life*. It was there that he began painting magazine covers. It was the start of an occupation that would continue for sixty-five years.

For forty-seven years, starting when he was only twenty-one, Rockwell became famous as a featured artist in the *Saturday Evening Post*. It all started thanks to cartoonist Clyde Forsythe, with whom the young Rockwell shared a studio in New Rochelle, New York. Forsythe helped Rockwell sell a cover called *Boy with Baby Carriage* to the *Post* in 1916. Rockwell painted 330 covers during his career.[4]

Arguably, Rockwell's most famous contribution to the *Post* was the Four Freedoms series, which was published in 1943 after seven months of work. Inspired by President Roosevelt's declaration that universal rights were built on four principles—the freedoms from want and fear, freedom of speech, and freedom to worship—the four paintings were exhibited nationwide by the War Department in order to promote War Bonds.[5]

Rockwell died on November 8, 1978, of emphysema at age eighty-four in Stockbridge, Massachusetts.

On his trip to Montmartre with Rockwell, Indiana encounters celebrated artist Pablo Picasso. He also meets other renowned artists of the time, including Edgar Degas, Georges Braque, and Henri Rousseau. It seems impossible that so many famous artists could be in one location at one time. Yet the real world is more unbelievable than anything imagined by television.

Pablo Picasso was a Spanish painter and sculptor. His full name was Pablo Diego José Francisco de Paula Juan Nepomuceno María de los Remedios Cipriano de la Santísima Trinidad Clito Ruiz y Picasso. One of the most acclaimed artists of the twentieth century, he was perhaps best known as the cofounder, with Georges Braque, of cubism.

Picasso was born in Málaga, Spain, the son of an artist who specialized in painting birds in natural settings. Picasso demonstrated astonishing skill at drawing from an early age. His father, beginning in 1890, acted as his first teacher. Picasso attended the Academy of Arts in Madrid but left after less than a year of study.

Picasso moved to Paris in 1900. He lived there with poet Max Jacob, who helped him to learn French. The two lived in extreme poverty, and Picasso burned much of his work just to stay warm. In 1901, he began to sign his work with only his last name, Picasso.

Over the next few years, Picasso worked in Barcelona and Paris. It was in Paris that he developed the art style known as cubism with Braque. By the end of the decade, Picasso was spending most of his time in the Montmartre artist section of Paris.

On Saturday, November 21, 1908, Picasso threw his notorious party for elderly artist Henri Rousseau, which Indiana Jones supposedly attended. A number of important artists came to the celebration and gave mock tributes to Rousseau. No one who attended seemed to know whether Picasso was serious in honoring Rousseau, or if he was throwing the party as one gigantic joke. No one ever found out, as Picasso never revealed his true motives. Picasso's girlfriend at the time made spaghetti for the guests, and everyone drank too much wine. Among those who attended the party was Gertrude Stein, who wrote an account of it years later in her autobiography.

Later in life, Picasso appeared in several films, always playing himself. Oddly enough, Picasso suffered from dyslexia. He died on April 8, 1973, at a dinner party. Picasso left no will, so his estate tax, which was owed to the French government, was paid off in paintings. These works became an important part of a Picasso museum in Paris.

Georges Braque, whom Indy supposedly met at Picasso's party, was a major twentieth-century French painter and sculptor who, along with Picasso, developed the art movement known as cubism. Braque began his career as an impressionist painter, but after seeing a display of art in 1905 by Henri Matisse and André Derain, who were known as the Fauves, he adopted that style. The Fauves believed in using brilliant colors and loose structures of form to inspire strong emotions.

From 1909 to 1914, when Braque joined the French army to fight in World War I, his interest in geometry and simultaneous perspective, as well as a close working relationship with Pablo Picasso, inspired work that questioned the fundamental standards of art. Their analytic cubism, as it would be called, was driven by their pulling objects apart and considering them in terms of their shapes. Later, their synthetic cubism would do the opposite, being based on pushing objects together. The former was characterized by neutral, monochromatic paintings of subjects faceted and divided; the latter, by their experiments with pasting pieces of paper on painting. Braque called this style, similar to collage, *papier collé*.[6]

Henri Julien Félix Rousseau was a self-taught French postimpressionist painter of primitive art. Most of his life, Rousseau was mocked by other artists for his style, but years after his death, he achieved recognition as a great artist. Born in the Loire Valley, Rousseau worked as a tax collector in Paris. He didn't start painting seriously until his early forties. His best-known paintings were of jungle scenes, although he never traveled outside France and never saw a jungle.

Rousseau's primitive style was ridiculed by most critics. Yet he never seemed aware of their criticism. He once told Picasso, "There are only two real artists in the world, you in 'Egyptian style' and I am in 'Classical.' "[7] Picasso's party was well timed, as Rousseau died less than two years later.

When Indy turns ten, he and his parents join ex-president Theodore Roosevelt on an African safari. Roosevelt's expedition is sponsored by the Smithsonian Institute; its mission is to bring home specimens of African wildlife for the National Museum in Washington, D.C.

Theodore Roosevelt, known to the public as Teddy, served as the twenty-sixth president of the United States. He was the youngest president, at forty-two, in the history of the United States. Teddy had previously served as governor of New York and was an explorer, an author, and a soldier. His was one of the four faces carved on Mt. Rushmore.

In 1901, while serving as vice president, he succeeded President William McKinley after McKinley was assassinated. As president, Teddy promoted the conservation of natural resources. He was known for attacking big business and earned a reputation as a "trust buster." He was the first American to be awarded the Nobel Peace Prize, in 1906, for negotiating the peace treaty in the Russo-Japanese War.

In March 1909, shortly after the end of his second term as president, Teddy went on safari in Africa. His mission was to obtain specimens for the Smithsonian and the Museum of Natural History in New York City. The safari was organized in part by Frederick Selous, the famous British big-game hunter and explorer, but Selous was not the leader of the expedition. This was the safari that Indy and his parents joined in Africa. The excessive killing of jungle creatures that took place on the safari disturbed young Indiana. The Roosevelt expedition killed or trapped more than 11,397 animals. Among the rare beasts slaughtered in the name of science were six rare white rhinos. The skins of the animals were shipped back to the Smithsonian, where it took years to stuff and mount them for exhibits.

Frederick Courteney Selous was famous for his exploits in Africa. He served as a model for H. Rider Haggard's hunter, Allan Quartermain. In World War I, Selous fought with the rank of captain in the 25th Royal Fusiliers in East Africa. He was sixty-five years old when he was killed by a German sniper's bullet.[8]

Indy next travels to Florence, Italy, with his parents and his tutor, Helen Seymour. While Indiana's father is away on business, Italian opera composer Giacomo Puccini takes the family under his wing. Indy finds Puccini fascinating until he realizes that the composer is more interested in Indy's mother than is considered proper.

Giacomo Antonio Domenico Michele Secondo Maria Puccini was an Italian composer whose greatest operas included *La Bohème*, *Tosca*, and *Madame Butterfly*. *La Bohème* is standard repertoire at houses such as the Metropolitan Opera in New York City; *Tosca* has one of the most provocative and exuberant plots in opera; and *Madame Butterfly* has exquisite soprano solos. Puccini was regarded by most opera critics as the successor to Giuseppe Verdi, also one of the most famous opera composers in the world to this day.

On a trip to Vienna, Indiana meets Princess Sophie at the Spanish Riding School. It's the first time Indy experiences the joys and frustrations of love. He's given advice on how to handle his emotions by three of the fathers of modern psychoanalysis—Freud, Jung, and Adler—but their words do nothing to calm his beating heart. He does everything he can to be with Sophie, but the romance is not to be—because she's the daughter of Archduke Franz Ferdinand, the ruler of the Austro-Hungarian Empire.

Indy again seems to encounter more prominent people than is believable—or possible. Yet Princess Sophie did vacation in Austria with her royal parents. Moreover, Freud, Jung, and Adler all lived in Vienna during this period. And the fathers of psychoanalysis met once a week to discuss the new science of the human mind they were formulating.

Neurologist Sigmund Freud remains the most famous founder of the psychoanalytic school of psychology. His name is a household word, and he's particularly well known for his theories about the unconscious mind, repression and defense mechanisms, sexual desire, and incorporating dreams as a part of therapy. Of course, he's most famous for spawning the phrase "Freudian slip." Although his theories have become more controversial and less accepted over time, they have nonetheless had a profound effect on American culture and how we look at art and ourselves.[9]

Carl Jung, a Swiss psychiatrist, was the founder of analytical psychology. Along with working as a psychologist, he spent much of his life exploring alternative thought concepts through philosophy, sociology, and literature. Jung felt that people believed too much in logic and science and not enough in spirituality and the unconscious.

An Austrian psychologist, Dr. Alfred Adler founded the school of individual psychology. In 1902, he was invited by Sigmund Freud to join a discussion group that met at Freud's home on Wednesday evenings. The group was the beginning of the psychoanalytic movement. These were the men who tried to help Indiana express his feelings for the princess. Adler remained a member of the group until 1911, when he formally left after disagreeing with Freud's theories. By this time, Freud and Adler had grown to dislike each other intensely. So much for the success of modern psychoanalysis in helping people to deal with personality problems.

Princess Sophie of Hohenberg was indeed a daughter of Archduke Franz Ferdinand of Austria and his wife, Sophie, Duchess of Hohenberg. Princess Sophie never ascended to the throne. In 1938, she and her two brothers were arrested by the Nazis and sent to the Dachau concentration camp. They were freed by Allied troops in 1945. The princess died in 1990.

Travels with Father: Three Writers, a Boy Genius, and a Famous Scientist

In Russia and now eleven years old, Indiana has a fight with his parents and runs away from home. He falls in with a renowned novelist named Leo Tolstoy and accompanies the writer on a journey through the Russian countryside. He argues philosophy with Tolstoy and encounters both gypsies and Cossacks. Indy, missing his parents, leaves Tolstoy and returns to his home.

Of course, Leo Nikolayevich Tolstoy was a real author. In fact, he was one of the world's finest novelists and is remembered for two masterpieces, *War and Peace* and *Anna Karenina*. Many literary critics consider *War and Peace* the greatest novel ever written.

As a philosopher, Tolstoy was highly regarded for his beliefs in nonviolent resistance. His ideas strongly influenced Mahatma Gandhi and Martin Luther King Jr.

Tolstoy went through a crisis in faith in his early fifties. He became a fervent believer in nonviolent, antigovernment pacifism. In his late seventies, Tolstoy became seriously ill and had to live in the Crimea for his health. The last month of his life was spent wandering through Russia with his daughter Alexandra. That's when he would have encountered a young Indiana Jones. Tolstoy died on November 7, 1910. Though it's possible that old Tolstoy would have met young Indiana, it's very improbable.

The Cossacks whom Tolstoy admired dwelled on the southern steppe of Eastern Europe and Russia. They were famous and they were feared. They were known for warfare and great bravery. Like many bloodthirsty tribesmen, they made good friends and terrible enemies. The Russian army apparently liked them, as the Cossacks served as military police, border guards, and general protectors of trading posts and settlements. In Napoleon's time, the Cossack's were so brutal and skilled that the French were more afraid of the Cossacks than of the enlisted Russian soldiers. The Cossacks considered Tolstoy one of their own and welcomed him as royalty whenever he visited.

The next stop for the Jones family is Greece, where Indy and his father ride a tiny cage up a thousand-foot-high mountain to visit a monastery. During this adventure, Indy and his father meet writer Nikos Kazantzákis.

Nikos Kazantzákis was the author of poems, novels, essays, plays, and travel books. He was the most important and most translated Greek writer and philosopher of the twentieth century. In 1964, a film by Michael Cacoyannis was released with the name *Zorba the Greek*. It was based on Kazantzákis's novel of the same name.

On his continued journeys, Indy passes through the Holy City of Benares, India. It's there that he meets and becomes friends with Jiddu Krishnamurti. Jiddu has been declared the New World Teacher of theosophy. As he struggles to be everything his worshippers want, he demonstrates to Indy the power of independent thought.

J. Krishnamurti was born in Madanapalle, India. Annie Besant and C.W. Leadbeater, the two leaders of the Theosophical Society, became convinced that Krishnamurti was the new incarnation of the

Maitreya Buddha, the so-called World Teacher. Unwilling to be labeled a living god, Krishnamurti publicly disavowed this notion and disbanded the multinational Order of the Star organization. He traveled all over the world lecturing that society could only be changed by individuals transforming themselves, since society was based on the interactions of individuals. He wrote a number of books and was awarded the 1984 UN Peace Medal. He died in 1986.

Indy's mother learns about faith when Indy gets typhoid fever while traveling through rural China, far from any modern doctor. She finds herself depending on poor Chinese villagers and their age-old treatments to save her son.

Indy then returns with his parents to the United States. His mother dies, but his father doesn't remarry. Father and son live for several years in Utah but then return to the East Coast.

Indy's adventures continue when he turns sixteen and is no longer under constant watch by his father or a tutor. His girlfriend, Nancy Stratemeyer, is the daughter of author and publisher Edward Stratemeyer. The young couple set off to explore Thomas Edison's lab, and while visiting the inventor, they must thwart deadly German spies who want for nefarious purposes Edison's electric automobile battery.

Again, while this adventure seems improbable, was it impossible?

Born in New Jersey, Edward Stratemeyer published and wrote children's books. The author of more than 150 novels, he created some of the most popular series of books for children ever published, including the Bobbsey Twins, Tom Swift, the Hardy Boys, and the Nancy Drew Mystery Stories.

Stratemeyer paid several freelance authors to write books as part of a long-running series under a house name owned by his company. Thus, the readers thought they were getting new novels by their favorite writer, featuring the same lead characters in book after book, while in reality the books were being composed by a number of different writers.

The novels produced by the Stratemeyer syndicate were different from most books being published for children and teens in the early years of the twentieth century, in that they didn't promote moral lessons or some sort of educational viewpoint. The books were written purely as entertainment and could be read in one sitting. The extremely popular books sold millions of copies. Needless to say, the novels were criticized by teachers and librarians at the time as not being of high moral caliber. Nowadays, teachers only wish their students would read as many books as they did in the 1920s and 1930s.

Nancy and Indy visit Menlo Park, New Jersey, about a half-hour's drive from Elizabeth, where they meet Thomas Alva Edison, perhaps the most famous American inventor of all time. Edison was known as "the Wizard of Menlo Park." Among his many notable discoveries were the long-lasting electric lightbulb and the phonograph. Some people credit Edison with developing the first scientific research facility.

Whenever Edison was considering hiring a new employee, he first took the man out for lunch. When the food was served, Edison watched to see whether the potential employee salted the meal before tasting it. If he did, Edison wouldn't hire the man. Food, Edison felt, needed to be tasted before it was seasoned, and he didn't want employees who acted without thinking.

It's not unbelievable to assume that German spies would be interested in Edison's inventions. Menlo Park is in New Jersey between Elizabeth and Princeton, so it's not outrageous to imagine that Indiana and Nancy might go there. Their adventure is unlikely but definitely not impossible.

Spring Break Adventure: Prelude to War

By this time, Indy's father realizes that his son enjoys adventures that often lead to danger, so to protect the boy, he sends Indy to New Mexico to live for a while with his aunt. But, of course, danger follows Indy. Pancho Villa kidnaps Indy, and the young adventurer finds himself in the throes of the Mexican Revolution, where he meets young army officer George Patton and General John "Black Jack" Pershing.

Somehow, Indiana fails to meet Ambrose Bierce, the noted American author who disappeared in Mexico hunting Pancho Villa. And at the time this episode takes place, the United States is not yet involved in the war.

Pancho Villa Doroteo Arango Arámbula, known as Pancho Villa, was one of the main leaders of the Mexican Revolution. Villa seized land from the wealthy and gave it to the poor. To pay for supplies, Villa robbed trains, moving quickly on the railroad systems to avoid detection. Many of Villa's tactics and strategies were later adopted by twentieth-century revolutionaries.

As a major revolutionary, Pancho Villa and his activities attracted journalists and even moviemakers from around the world. Despite much research into his life, many of the facts about his career are still in dispute.

Villa was perhaps the first revolutionary to realize the importance of good publicity in fighting a war. He made a deal for $25,000 with the Mutual Film Company of New York, giving it exclusive film rights to all of his battles. One of the mostly widely seen newsreels in the early years of the twentieth century consisted of footage of these engagements.

Pancho Villa was an early version of the flamboyant and egocentric movie star of today. He staged scenes for his movies. He reenacted real fights that were long over. He even insisted that his hair be done to make him look more handsome. And because he was also a warmonger, he delayed an attack on Juarez so his exploits, rather than the World Series, which was taking place at that time, would be center stage in newspapers.[10]

Villa, angry because the U.S. government refused to support his revolution, raided Columbus, New Mexico, on March 9, 1916. His forces killed sixteen members of the U.S. 13th Cavalry Regiment stationed in the town, as well as several businessmen. Villa's men seized several machine-guns and some ammunition and then burned the town before retreating to Mexico. It was the only successful military attack that took place in the continental United States in the twentieth century.

Ten days later, President Wilson ordered ten thousand soldiers into Mexico under the command of General John J. Pershing, to find and capture Pancho Villa. Serving under Pershing was Lieutenant George S. Patton.

It's worth mentioning that Patton, who served with distinction as a general in World War II, was an exceptional athlete as a young man. He participated in the 1912 Olympics on the U.S. team and placed fifth in the pentathlon, which was won by Jim Thorpe. After the Olympics, Patton was the army's youngest-ever "Master of the Sword."

In June, Indy witnesses a barroom showdown between Julio Cardenas, Pancho Villa's personal bodyguard, and George Patton. Patton kills Cardenas and carves a notch in his Colt Peacemaker revolver. Pershing nicknames Patton his "Bandito," and the gun battle is written up in newspapers across the United States, burnishing Patton's reputation in the military.

Indiana, no longer a prisoner of Villa, decides to accompany his Belgian friend Remy to Europe to fight in World War I.

Indiana doesn't realize it at the time, but when he travels to Europe to fight in the war, he will be followed by Pershing and Patton a little more than a year later when the United States formally enters the war.

When Pershing was made commander in chief of the American Expeditionary Force (AEF), the United States army was made up of, at most, twenty-five thousand troops. Over a period of a year and a half, the army grew to nearly three million men. Pershing personally led the successful Meuse-Argonne offensive of 1918. He was considered a mentor to an entire generation of American generals who served in Europe during World War II, including George C. Marshall, Dwight D. Eisenhower, Omar Bradley, and George S. Patton.

Directly before the Easter Rebellion, Indiana is in Ireland, and in the midst of revolution and politics. He also falls in love again, and through his girlfriend's brother is involved with the Irish resistance. He also meets poet William Butler Yeats, author Sean O'Casey, and Irish revolutionary Sean Lemass. It's another case of Indiana being in the right location at the right time.

William Butler Yeats was born in Dublin, the son of a well-known portrait painter. Along with Edward Martyn and Isabella Augusta,

Lady Gregory, Yeats founded the Irish Theatre, which later became the Abbey Theatre. He served for many years as its lead playwright. His plays usually dealt with Irish mysticism and legends.

After 1910, Yeats's work became more esoteric and was aimed at a small literary audience. In 1922, he was appointed to the Irish Senate. He won the Nobel Prize in Literature in 1923 for his poetry.[11] After winning the Nobel Prize, he went on to compose some of his finest works. His later volumes of poetry established him as one of the most influential twentieth-century writers.

Sean O'Casey was a Dubliner who wrote three realistic plays about life in the Dublin slums that were all performed at the Abbey Theatre. In 1929, O'Casey submitted a new play to the theater about the horrors of World War I. It was rejected, and O'Casey never submitted any more of his work to that theater.

Sean Lemass was an Irish revolutionary who helped to form the Fianna Fáil political party in 1926. He became a practical, pragmatic politician in order to best serve his country. In 1959, he became the premier of Ireland and helped to modernize his country's industry. He retired in 1969 and died in 1971.

That Indiana was able to meet three of the most outstanding men in modern Irish history over the course of a few weeks' stay in Ireland seems improbable. Still, history has a way of playing tricks with notoriety. All three men lived in Dublin during the time Indy was there. Circulating among the intellectuals and the revolutionaries of the time gave Indy the opportunity to socialize with a number of people who were destined for greatness. What is often forgotten in the emphasis on whom Indy meets in his adventures are the many people he encounters who are startlingly ordinary.

Indiana seems to get into a lot of trouble when he falls in love. This time, he goes to England and becomes involved with a young woman who wants her independence. Moreover, she is heavily involved with the women's suffrage movement. Indy ends up fighting in the streets over women's rights.

Indy's latest girlfriend qualifies as a suffragette, meaning "somebody who believes in the right to vote." In England during the early

1900s, suffragettes believed in direct action to obtain the right for women to vote. To make their point, the suffragettes marched, and they also took more drastic measures, such as smashing windows and setting mailboxes on fire. In rare instances, they even set off bombs. Many suffragettes were imprisoned, where they conducted hunger strikes.

Women campaigned for the vote on both sides of the Atlantic. Suffragettes in the United States produced posters of Woodrow Wilson, referring to him as "Kaiser Wilson."

World War I led to a serious shortage of manpower in both the United States and Britain. Jobs that had always been handled by men were suddenly being done by women. The notion that women were capable only of cooking, cleaning, and raising children was proved totally untrue. It became clear to just about everyone that women could perform as well as men in the workplace.

In 1918, the British Parliament passed an act granting the right to vote to some but not all women. In the United States, women's right to vote became law with the passage of the Nineteenth Amendment to the Constitution in 1920. All women in the United Kingdom were finally granted the vote in 1928. Thus Indiana, who witnessed the rise of the suffragette movement, saw its goals come to pass.

While in England, Indiana witnesses a zeppelin air attack on London, then attends an oddball dinner party with Winston Churchill.

With images of the Goodyear blimp in mind, it's hard to believe that zeppelins could ever have inspired terror among the population of England. At a time when airplanes were still a novelty, however, few people realized that lighter-than-air vessels could be used for warfare. The German High Command did, and they invented aerial warfare. Indy's adventure in one of the first air raids was just a hint of the future, right out of the pages of an H. G. Wells novel.

In 1914, the Germans created the war zeppelin, which could fly more than two miles high at almost ninety miles an hour. Onboard the zeppelin were four thousand pounds of bombs and many machine guns.[12]

In January 1915, the surprising power of the zeppelins was brought home to England when two warships bombed Yarmouth and King's Lynn. When London was attacked on May 31, there were eighty-eight victims, with twenty-eight killed. The next two years would see more of the same, as more than five hundred British civilians were killed in air raids across the country.[13]

The English, though, are a resourceful people, and they soon figured out how to meet the challenge of the zeppelins. Zeppelins could fly, but not particularly well. They performed very poorly in bad weather, and they did even worse when confronted with highly motivated British fighter pilots, who soon became experts at destroying the clumsy aircrafts. All the pilots had to do was shoot the balloons. Of 115 zeppelins used in the war, 77 were either destroyed or grounded, and in 1917 the air raids ceased.[14]

The Right Honorable Sir Winston Leonard Spencer Churchill was the son of Lord Randolph Churchill and an American mother. He was educated at Harrow and Sandhurst. After a brief but eventful career in the Boer War, he became a Conservative member of Parliament in 1900. He held many high posts in Liberal and Conservative governments during the first three decades of the century. He was a strong voice warning of the coming Nazi peril but was ignored until it was almost too late.

At the outbreak of World War II, Churchill was appointed First Lord of the Admiralty, a post that he had earlier held from 1911 to 1915. In May 1940, he became prime minister and minister of defense. Churchill remained in office until 1945. He took over the premiership again in the Conservative victory of 1951 and resigned in 1955. He remained a member of Parliament until the general election of 1964, however, at which time he did not seek reelection. Queen Elizabeth II made Churchill a knight. President Kennedy granted him an honorary citizenship of the United States in 1963.

Churchill was a prolific author and wrote numerous books on British history, especially covering the years in which he was a major participant. He was awarded the Nobel Prize in Literature in 1953. Churchill was noted for his acerbic wit. The following exchange has been reported so many times, it must be true.

Churchill, drunk at a party, runs into an old-fashioned dowager who disapproves of liquor. "You, sir," she says, "are drunk."

"Yes, ma'am," replies Churchill. "But you are ugly. Tomorrow, I will be sober, and you will still be ugly."

It's worth mentioning that on Indiana's short trip to the United Kingdom, he met two future Nobel Prize in Literature award winners, Churchill and Yeats. Of course, Indy had already met one Nobel Peace Prize winner when he was only ten—Teddy Roosevelt.

Trenches of Hell: The Horrors of War

Indiana lies about his age and joins the Belgian army to combat the Germans. While fighting in the trenches, Indy participates in one of the bloodiest conflicts of all time, the Battle of the Somme. After experiencing the horrors of trench warfare and gas attacks, Indy is taken prisoner by the Germans and put in a POW camp.

Indy's experiences in World War I change an idealistic youngster into a cynical young man. War, as shown onscreen and written about in many books, is a dangerous enterprise but also an opportunity for men to become heroes. Yet after life in the trenches on the front lines, Indiana no longer wants to be a hero. He wants to remain alive. What terrible events turned this idealistic young boy into a hardboiled man?

Indy arrived in Europe just in time to participate in the Battle of the Somme, one of World War I's longest and most bloody battles. More than a million men died in the battle as the Allies tried to distract the Germans from the Battle of Verdun. The bloodbath took place along the Somme River in northern France.

In our modern times, when the slaughter of a dozen men during one week of fighting makes headlines, the thought of thousands dying in twenty-four hours is unthinkable. On July 1, 1916, the first day of the Battle of the Somme, more than nineteen thousand British soldiers died. Actual footage showing the deaths of many appeared in *The Battle of the Somme*, a film released that August.

The British Expeditionary Force was unable to gain an advantage following July 1, and the French only advanced slightly into German-occupied areas.[15] The British and French generals never wavered in their belief that the Germans were on the point of surrender and that sooner or later the Allied forces would achieve victory. Thus the offensive continued throughout the summer and into November. Poor weather—snow—brought a halt to the Somme offensive on November 18. During the month's-long attack, the British and French armies had advanced less than ten miles, at a cost of 420,000 British casualties. The French lost more than 200,000 soldiers. German casualties were estimated at 500,000 men. It was Indy's first taste of modern warfare, and he quickly realized that he didn't like it. There was no romance in being a soldier, only death and the trenches.

Nothing in war is entirely new. Both the Civil War of 1861–1865 and the Russo-Japanese War had trench warfare, but the method reached its horrific potential during World War I.

Trench warfare consisted of two opposing systems of trenches fairly close to each other, in which opposing armed forces attacked, counterattacked, and defended themselves while operating from relatively permanent systems of trenches dug into the earth at or below ground level. With machine guns prominent in war, trenches became critical for protecting soldiers. The men could no longer move across the battlefield from the trenches because of the danger of the machine guns. Hence, soldiers remained in trenches, and battles became stalemates.[16] To break through enemy lines, soldiers tossed bombs into trenches, used mustard gas, and eventually used tanks.

In World War I, a series of outflanking movements, known as the Race to the Sea, led to trench lines being dug from the Belgian coast to the Swiss frontier as both sides tried to protect themselves from artillery and machine-guns. These weapons also caused the Allied campaign against Turkish forces at Gallipoli, from April 1915 to January 1916, to degenerate into trench warfare. The trench system along the western front ran for approximately 475 miles, in an *S* shape across Europe, from the North Sea to Switzerland.[17]

There were many types of trenches, including those used for cover, for storing supplies, for firing at enemies, and for communications work. Cover trenches protected soldiers in the firing trenches. Often, the men lived in the trenches.

The trenches themselves were their own sort of no-man's-land, a Hobbesian world of continual fear and danger from violent death where a soldier's life was, if not solitary, still poor, nasty, brutish, and likely short. At any moment shells might fall from above or a sniper might take aim from across the lines. The trenches were frequently filled will water and mud, crawling with lice and rats, and open to the worst of the elements; they were vast cloacae of disease and despair.[18]

Eventually, trench warfare wasn't critical to battles, as the machinery of war developed. Trenches were no help against rocket launchers, advanced tanks, and air attacks. Science and technology often seem to progress the most during wars, and the advancements take their toll on human lives.

After witnessing the Battle of the Somme, Indiana no doubt thought that he had witnessed the worse that war could offer. He was wrong. World War I was the first war in which poison gas was used on the battlefield. For a young man like Indy, far from home and stuck in the trenches, poison gas attacks must have seemed like a preview of hell. Which side used poison gas? Did it help the Allies to win the war?

In 1914, the French launched the first gas attack used in warfare. Fighting the Germans in World War I, they filled tear-gas grenades with xylyl bromide and tried to stop the German advance. While the gas was irritating to the eyes, it did not poison anyone. It was used in desperation as the German army seemed invincible.

Later in 1914, the Germans retaliated when they attacked the French at Neuve Chapelle using gas shells that made the French sneeze violently. Again, the gas did not poison or kill anyone.

Both of the gas attacks occurred before trench warfare became pronounced in the war. In the trenches, the war sunk into a stalemate, so the leaders used poison gas in hopes of gaining an advantage. They hoped that the gas would cause widespread panic, and, of course, they figured the poison gas would kill vast numbers of enemy soldiers. They also figured that using the gas might bring the war to an end much more quickly.

In the late afternoon of April 22, 1915, at the second Battle of Ypres, the Germans used chlorine gas for the first time. From pressurized cylinders planted before their lines, a vast yellow-green cloud sprang and bloomed and flowed toward the French lines. The French, who thought it might be just a smokescreen to hide German troops, held their positions, and indeed the Germans were supposed to follow the gas cloud into battle. But when the gas reached its target, the French and Algerian soldiers fled screaming. The German soldiers were balked by their terror and failed to finish the attack.[19]

After Ypres, every other nation felt that since the Germans had used the poison gas, they could also use it. Once the rules of engagement were broken, there was no going back.

The first Allied use of poison gas took place in 1915. The British attacked the German trenches at Loos in September. Following the lead of the Germans, the British used pressurized gas containers to hold the poisonous chlorine gas. At the time, the wind was blowing toward the Germans. Once the gas was released, however, the wind changed direction and starting blowing toward the British. More than two thousand British soldiers were injured, and seven died.

Looking for something even worse than chlorine gas, scientists working on both sides of the war came up with chemical formulas for more deadly gases. Phosgene gas was especially dangerous because its impact wasn't felt until forty-eight hours after it had been inhaled, and by then it was already doing serious harm to the lungs.

Mustard gas, which the Germans introduced against the Russians in September 1917 at Riga, seemed to marry the worst traits of chlorine and phosgene by causing both internal and external blisters within hours of exposure. It did not always kill, but it never failed to cause intense pain and often blindness.[20]

The number of soldiers actually killed by poison gas during World War I was relatively small. More than 180,000 British soldiers were harmed by gas, but only 8,000 of them died. Altogether, approximately 91,000 men died from gas attacks, but more than a million of them were listed as gas casualties. Yet no follow-up studies were ever conducted by the Allies to see how many of the men affected by gas died years later from their injuries.

Nonetheless, poison gas remained the war's most feared weapon. A soldier could take shelter from gunfire. He could have a bullet removed and be healed. He knew there was another man wielding that other gun, a man who could be defeated. But gas could not be fought. Gas was silent and indiscriminate. Gas came when there was no attack. And gas left its victims in agony for no sure amount of time before they succumbed. It tortured the mind and the body.[21]

When thrown in a POW camp, Indy finds himself plotting with a young Frenchman named Charles de Gaulle. They plan a daring escape.

Charles De Gaulle became a French military leader and statesman. Charles André Joseph Marie de Gaulle was born in November 22, 1890, and raised in Paris. De Gaulle attended military school as a young man. After graduating in 1912, he decided to join an infantry regiment rather than an elite corps.

During World War I, Captain de Gaulle was severely wounded at the Battle of Verdun and was left for dead on the battlefield. He was found and taken prisoner by the Germans. During the next two years, he tried to escape five times. He finally was freed when the war came to an end.

During the 1930s, de Gaulle wrote numerous articles attacking France's reliance on the Maginot Line as protection from a German invasion. He strongly favored reliance on a mechanized armored force for protection. His advice was totally ignored, and in June 1940, the Germans easily swept past the Maginot Line and conquered France. De Gaulle established a French government in exile in London. He became a leader of the Free French.

After the liberation of Paris in August 1944, de Gaulle received a hero's welcome in the city. He served as president of the French Republic several times. On November 9, 1970, he died of a heart attack.

Once Indiana escapes the POW camp, he acts as a messenger for the French generals in charge of the war. Disgusted by their win-at-any-cost attitude, he goes to Paris, where he has a brief fling with Mata Hari. Soon after, Mata Hari is arrested and accused of spying for the Germans.

Mata Hari's real name was Margaret Gertrud Zeller. She was born in Java in 1876 to a Javanese mother and a Dutch father. When she was a teenager, she moved to Holland. A fine linguist, she was also a noted conversationalist. She was both intelligent and beautiful.

After changing her name to Mata Hari, which meant "Eye of the Morning," Margaret moved to Paris, where she worked as an exotic dancer. Of course, many important men, including French officers, wanted to bed her, or at least enjoy her favors. She was still in Paris when the war broke out in 1914. In 1916, she was arrested by the French and accused of being a spy for the Germans.

Evidence against Mata Hari was mostly circumstantial, but nonetheless she was sentenced to death and executed by firing squad in 1917, when she was only forty-one years old.

Assuming that Indiana had a short affair with Mata Hari in 1916, as dramatized in his adventures, he would have been sixteen or seventeen years old at the time. She would have been forty, at least as old as his mother. Sometimes fact checkers need to verify their facts more closely.

The Phantom Train of Doom: At War in Africa

The Allies send Indiana to destroy a very mysterious German artillery gun. This gun can somehow become invisible, then reappear, both without warning or apparent cause. Indiana is accompanied on his adventure by a group of old and reckless soldiers, and after a harrowing journey through Africa, they come across a hideout in the mountains. They discover that the Germans have been moving the mysterious disappearing and reappearing gun by railroad through the jungle.

Most people don't know that World War I was fought in Africa as well as in Europe. The battles Indy witnesses are part of forgotten history. Apart from the combat making entertaining TV fare, we have to ask whether these engagements had any influence on the war in Europe. Did the men fighting them become famous or are they all forgotten? These battles are true mysteries of the Indiana Jones shows.

In Indy's months in Africa, he meets Albert Schweitzer, Colonel Paul von Lettow-Vorbeck, Richard Meinertzhagen, and Jan Christian Smuts. Other than Schweitzer, were any of these men known for anything they did other than fighting in World War I? Are they names that deserve greater recognition in world history?

Colonel Paul Emil von Lettow-Vorbeck, a German general in World War I, was never defeated. He fought with an army of approximately fourteen thousand soldiers plus eleven thousand African natives and a few thousand Germans. An expert at guerrilla warfare, he was able to fend off British and South African armies that were huge, often a dozen or so times larger than his army.

Over several years, Lettow-Vorbeck raided the British colonies of Kenya and Rhodesia, destroying military outposts and railroad stations. In March 1916, Jan Smuts and an army of 45,000 men launched an attack on Lettow-Vorbeck from South Africa. Despite overwhelming odds, Smuts was unable to catch the German commander or put a halt to his raids.

In 1918, Lettow-Vorbeck learned of the end of the war on November 11 from a British prisoner. Lettow-Vorbeck, an honorable soldier, surrendered to the British on November 25, 1918, at Mbaala, Zambia.

Lettow-Vorbeck returned to Germany as a national hero. He turned to politics and served in the Reichstag from 1920 to 1930. His memoirs of his wartime adventures were published in book form as *My Reminiscences of East Africa*. He died on March 9, 1964.

Richard Meinertzhagen was born in 1878 in London. In January 1899, he was commissioned into the royal fusiliers, and over time he was stationed in India, Africa, and Palestine. He served in Kenya and East Africa.

In World War I, Meinertzhagen was an intelligence officer for the Tanga expeditionary force that fought Paul Lettow-Vorbeck. In

October 1917, the Ottoman military was able to obtain British battle plans that Meinertzhagen supposedly lost, and the Turks tried to
dodge a planned attack, but the British ending up attacking them
from somewhere else. The plans were false. Meinertzhagen lost
them on purpose.

Meinertzhagen attended the 1919 Paris Peace Conference, and
as Edmund Allenby's chief political officer, he helped create the
Palestine Mandate. His work with the mandate supported the formation of Israel. Although he attained the rank of colonel, he was
dismissed from the service for insubordination in 1926. He spent
the rest of his life traveling in Asia.

Jan Christian Smuts had a significant career as a South African
and British Commonwealth statesman, holding various cabinet ap-
pointments as well as serving as prime minister of the Union of
South Africa twice, from 1919 until 1924, then from 1939 until
1948. He was also a prominent military leader, serving as British
field marshal in both world wars.[22]

During the Second Boer War, Smuts led commandos, and during World War I, he led South African armies against the Germans
and also the British Army in East Africa. He was instrumental in the
creation of the Royal Air Force, being a member of the British War
Cabinet from 1917 through 1919. He also served in Winston
Churchill's war cabinet in World War II. His signature was on the
treaties that ended both world wars.

Following World War I, Smuts helped create the League
of Nations, and years later he also helped establish the United
Nations. In fact, he wrote the preamble to the United Nations
Charter. He signed the charters of both the United Nations
and the League of Nations, and he was a strong supporter of
Israel.

*After destroying the train, Indy and his group of old, reckless soldiers
(aka "the Old and the Bold") are sent on a new mission, to capture
German general von Lettow-Vorbeck. During the adventure, Indy
ends up in a small village in Africa, where disease is rampant and*

children are dying. He is able to save one child, but it almost costs him the secret mission that is so crucial to the Allies. Does he sacrifice an innocent child for the good of his country? A very confused and unhappy Indy encounters Albert Schweitzer in the jungle. Indy finds his faith restored when he is able to leave the child to be healed at Schweitzer's hospital.

Albert Schweitzer was born in the territory known as Alsace-Lorraine, which at the time was part of the German empire. He was multitalented, and in addition to being a physician, he was also a musician, a philosopher, and a theologian. From 1901 to 1912, he held administrative positions at the Theological College of St. Thomas. The publication in 1906 of his book *The Quest of the Historical Jesus* established him as a major theologian.

During this period, Schweitzer gained fame as a concert organist. Having started to play when he was just a child, he soon was recognized as one of the finest organ players in the world. The money he earned from his recitals helped to pay for his continuing education and supported his later charitable work. Schweitzer even wrote a book on organ building and playing.

In 1906, Schweitzer decided that he wanted to go to Africa as a medical missionary. He thus began to study medicine at the University of Strasbourg, earning his degree in 1913. He then went to French Equatorial Africa and established a hospital at Lambaréné. In 1917, he and his wife were sent to a French prison camp, and they were released in 1918. Schweitzer spent the next six years in Europe, raising money for the hospital by performing in concerts, giving lectures, and writing books.

In 1924, Schweitzer returned to Lambaréné and expanded the hospital to seventy buildings. He was awarded the 1952 Nobel Peace Prize, and true to nature, he used the $33,000 prize to start a treatment center for lepers. He died on September 4, 1965.

Thus, Indiana met two Nobel Peace Prize winners while he was in Africa, Albert Schweitzer and Teddy Roosevelt.

Attack of the Hawkmen: Working with the French Secret Service

Indy ends up working for the French Secret Service, who send him on his next assignment, with the Lafayette Escadrille flying unit. He must perform deadly missions behind enemy lines. An encounter with German Ace Manfred von Richthofen leads to a battle that Indy cannot win, and he ends up on the ground behind enemy lines. Learning that there's a new, highly advanced secret weapon in the hands of the Germans, Indy sets off to find its inventor, German aircraft designer Anthony Fokker.

Just as the Royal Air Force would have its Eagle Squadrons in World War II, the French Air Service had its Lafayette Escadrille in World War I, a unit made up largely of volunteer American pilots who wanted to get into a war and fight for a cause to which their country hadn't yet committed. It was formed by the director of the American Ambulance Service, Dr. Edmund L. Bros, and an American expatriate already flying for France, Norman Prince, who wanted to fire up the enthusiasm of their countrymen to force the government's hand.[23]

Before long, the squadron was transferred to Bar-le-Duc, which was closer to the front lines. The original name of the group seemed to connote that the United States and France were allies, so with the Germans in an uproar, the United States changed the name.

The Battle of Verdun was the first test of the squadron's strength, and they battled at the front lines in 1916 for approximately five straight months. The squadron had a core of thirty-eight pilots, and when these men suffered deaths and serious injuries, other American pilots quickly took their place. In fact, there was a demand from pilots to join this elite squadron, and a special Lafayette Flying Corps made up of additional volunteers supplemented the air squad. Of the 265 Americans in the corps, 63 died during the war. The core squadron lost nine men overall.[24]

German fighter pilot Manfred Albrecht Freiherr von Richthofen was the most accomplished flying ace in World War I and was responsible for eighty combat wins. He had many French nicknames, among them "le Diable Rouge" (the Red Devil) and "le Petit Rouge" (Little Red). The English had nicknames for him too, such as "Red Baron" and "Red Knight".

Richthofen was originally a cavalry scout during World War I, but cavalry operations dwindled as the use of machine guns rose. Dissatisfied simply to participate in rote operations on the battlefield, Richthofen transferred to the Flying Service in May 1915.[25]

Anxious to be of service, Richthofen was frustrated further in June of that year when he was forced to begin the required three months of training. He assumed that by the time he finished his training the war would be over, and he would never be able to contribute to his country's success.

When in 1915, the Germans could no longer sustain their ground efforts against the Russians, they sent Richthofen to Ostend, where he became a fighter pilot. His first training flight was in October, and by the end of 1915, he passed his exams and officially became a pilot.[26]

Nine months later, Richthofen shot down his first plane, and to honor the event, he requested that a jeweler in Berlin make a special silver cup engraved with the date of his flight and the type of airplane he had destroyed. Being an ace pilot, he eventually accumulated sixty trophy cups in this manner.[27]

It didn't take long for Richthofen to earn the Order pour le Merite and to be appointed commander of the "Eleventh Chasing Squadron." As the supreme leader, he proudly painted his Fokker Albatros a bright shade of red. All members of his squadron also painted their fighter planes bright red, but unlike Richthofen's all-red Fokker Albatros, the other planes had to use additional colors to supplement the red paint.

In total, Richthofen destroyed 80 planes, and he was responsible for killing more than 125 American, English, and French pilots and engineers. In 1918, Richthofen was flying over Morlancourt Ridge near the Somme River when a bullet killed him.

Anthony Fokker, a Dutchman, was never interested in school and never completed high school. When Fokker was twenty years old, his father sent him to Germany to learn how to become an auto mechanic. Anthony was interested in flying, however, and instead studied airplanes. He soon earned his pilot license and built his own airplane. He became famous in the Netherlands by flying around the tower of the Sint-Bavokerk in Haarlem on August 31, 1911.

In 1912, Fokker opened his first company, Fokker Aeroplanbau, but soon he moved the factory to Schwerin and renamed it Fokker Flugzeugwerke GmbH. Later, he shortened the factory name to Fokker Werke GmbH.

As soon as the war erupted, Fokker lost his business to the German government, though he did continue designing airplanes and functioning as the director of his former company. His Fokker Dr. I became famous, thanks to Manfred von Richthofen, and the company eventually provided the Germans with close to seven hundred fighter planes.[28]

After the Treaty of Versailles was signed, the German government was no longer allowed to build airplanes, so Fokker went to the Netherlands. He started building civilian planes for his new company, the Dutch Aircraft Factory. In 1922, he founded yet another company, the Atlantic Aircraft Corporation in the United States, and eventually he became a U.S. citizen. He died at age forty-nine in New York City.[29]

Indy sets off on another dangerous mission in hopes of ending the war, this time to the palace of Austria's Emperor Karl. To get there, he must go undercover through enemy territory with the help of Hapsburg royalty.

In the middle of the Russian Revolution, Indy is sent to Russia on another dangerous adventure. Undercover, he joins up with some young Bolsheviks, only to find that he understands their dilemmas and inability to choose their own futures. He must remain faithful to his military obligations, yet he also feels obligated to help his friends.

A man for any task, Indy then goes undercover as a dancer in the Ballets Russes. While in Spain and dancing in the ballet, Indy encounters Pablo Picasso, with whom he was friends some time ago. Indy's mission is to discredit German diplomats, but he spends a lot of his time with three comic spies on various humorous adventures.

All Indy has to do is install a phone and await instructions, but he cannot seem to complete his mission. Paperwork and rules thwart his every attempt. Franz Kafka offers to help him, and Indy once again is disillusioned about the absurdity and horror of war.

Who were these people whom Indiana met in his spy days? And what was their claim to fame?

Karl Franz Josef Ludwig Hubert Georg Maria von Habsburg-Lothringen, more commonly known as Emperor Karl of Austria, did not live for long. He was born in 1887 and served as the last emperor of Austria. He also was the last king of Hungary and Bohemia and the nephew of Archduke Francis Ferdinand, whose assassination was the catalyst that started World War I. In 1918, Austria became a republic, and the emperor predictably refused to accept the new government. He died in 1922 after fruitlessly trying to restore his title for years.

The Jewish Franz Kafka was one of the most famous and respected writers ever to live. Most of his writing was published after he died, and it included amazing works such as "The Metamorphosis" and "The Judgment." He is most famous for merging surrealism and ordinary bleakness in his fiction.

Kafka, educated as a lawyer, also invented the safety helmet for which he received a medal in 1912. His helmet dramatically reduced deaths in Bohemian steel mills.[30]

After suffering through many unsuccessful tuberculosis treatments, Kafka checked into a Vienna sanitarium, where he could not even eat due to the condition of his throat. In 1924, Kafka died of starvation.

In 1903, at the Second Party Congress, various groups split from one another, including the Bolsheviks, who had been part of the Marxist Russian Social Democratic Labour Party. During the Russian Revolution, the Bolsheviks seized power, and the Soviet Union was born, along with the Communist Party.[31]

Vladimir Lenin, the leader of the October Revolution phase of the 1917 Russian Revolution, was the founder of the Bolsheviks. The revolutionaries did not cooperate with other parties, all of whom they called "the bourgeois," and their chief objective was to overthrow the czar.

Daredevils of the Desert: On the Edge of Africa

Next, Indiana helps the British attack Beersheba in the desert of the Middle East. With a gorgeous female spy, who also happens to be an accomplished and sexy belly dancer, Indy manages to save Beersheba from explosives that the Turks have implanted in water wells throughout the town. The Australian Light Horse Brigade charges to the rescue at the end.

Today, the city of Beersheba is the largest in Israel's Negev desert. It is no longer a desert *town*. In fact, by 2005 it was considered the sixth largest *city* in the country. The famous Ben-Gurion University as well as other prominent institutions, including the Soroka Medical Center, are located in Beersheba.

Since 1948, when Israel formally became an independent nation, Beersheba has grown in leaps and bounds. Its population consists of Jews from Arab countries as well as Jews from the former Soviet Union and Ethiopia.

In biblical times in Israel, there was no city or town farther south than Beersheba. Archaeologists estimate that Beersheba has existed since the year 4 BC. During the Sinai and Palestine Campaign in World War I, specifically on October 31, 1917, Beersheba played a pivotal and crucial role when the Australian 4th Light Horse Brigade defeated the Turks and captured Beersheba's wells. Many historians peg this as the last successful cavalry charge in history.

To preclude surprise attacks, the troops of the Brigade were dispersed in different locations. Although it took more than an hour to assemble the cavalry, the scattering of the soldiers turned out to be a wise strategy. They didn't have swords, so the soldiers flashed their bayonets in the sun so that the Turks would think the bayonets were swords.[32]

The Turks, completely surprised by the charge, failed to alter the sights on their rifles and ended up firing over the heads of the riders. Most of the Turkish soldiers surrendered, though a few stood and fought. When the main trenches before Beersheba were reached, some horsemen dismounted to capture the enemy with rifle and bayonet. Others raced through to the town to seize important objectives. The actual battle was over in minutes, the shocked Turks defeated and the equally astonished Australians the victors. It was one of the most exciting charges in the history of warfare, and Indy was lucky to be a witness to it.

Once again, Indiana wants to help bring the war to an end as quickly as possible. This time he's in Italy behind enemy lines and finds himself competing with none other than Ernest Hemingway for a beautiful girl.

Born in 1899 in Illinois, Ernest Miller Hemingway graduated from high school in 1917 and worked at a newspaper for six months, where he learned to use short paragraphs and short sentences. The method stuck with him for life.[33]

In 1918, Hemingway drove an ambulance for the Red Cross on the Italian front. It was during this period that he would have run into young Indiana Jones. On July 8, 1918, an Austrian trench mortar shell hit Hemingway while he was delivering supplies to the army. He eventually settled in Chicago and married Hadley Richardson, but shortly after the wedding the couple moved to Paris.

In Paris, Hemingway met Gertrude Stein, who introduced him to the "Parisian Modern Movement" in Montparnasse, and she became his first literary mentor. Ezra Pound became his other literary mentor.

In Our Time, Hemingway's first book, was published in 1925. The book introduced Nick Adams, a character who was in many stories by Hemingway. In 1926, Hemingway's first novel was published. It was called *The Sun Also Rises*.

In 1927, Hemingway divorced his wife, and with his new wife, Pauline Pfeiffer, he moved to Key West, Florida, in 1928. In 1929, *A Farewell to Arms* was published, and Hemingway made so much money from this one book that he was financially independent.

Over the next decade, Hemingway divided his time between Key West, Africa, and Spain. He reported on the Spanish Civil War and used his experiences in Spain as the basis for his novel *For Whom the Bell Tolls*, which was published in 1940. The same year, he divorced Pauline. A few weeks after the divorce was final, Hemingway married his companion of four years in Spain, Martha Gellhorn. After four years of constant arguments, he divorced Martha and married Mary Welsh.

In 1953, Hemingway won the Pulitzer Prize for his short novel *The Old Man and the Sea*. In 1954, he was awarded the Nobel Prize in Literature. Suffering from a number of ailments, including serious depression, Hemingway committed suicide on July 2, 1961, in Ketchum, Idaho.

Indy joins the French Foreign Legion and goes to North Africa for his next mission, which is to determine the identity of a traitor. He fights Berber tribesmen, and along the way falls for another woman, this time author Edith Wharton.

The French Army created the French Foreign Legion in 1831. Because foreigners were not allowed to be in the army, the Legion was established so these other soldiers could enlist and help the French protect its colonial empire. But the Legion went on to do much more, and ended up fighting against other countries in all of France's wars.[34]

The French Foreign Legion has been around for a very long time. Its members fought in both world wars, lasting through the destruction of the French colonial empire. It even survived and continued on when the French lost Algeria, where the Legion was first established.[35]

In 1926, a book by P. C. Wren called *Beau Geste* was published. The novel was about the French Foreign Legion, and it was adapted into many films, all of which boosted the fame of the group.

Edith Jones (Wharton) was born in New York City on January 24, 1862, into one of New York's wealthiest families. She was educated at home and overseas by a governess, and she loved reading. While still a teen, she began to write poetry, some of which was published anonymously in the *Atlantic Monthly* in 1880. At twenty-three, Edith married a wealthy Boston banker, Edward Wharton. They moved to Newport, Rhode Island, and named their home "Land's End."

Unhappy with the ugly exterior of her home, Edith decided to decorate the interior to express her personality. Working with architect Odgen Codman, she wrote a how-to book titled *The Decoration of Houses*, which was quite popular.

Edith, bored with her drab social life, turned to writing. She moved to Lenox, Massachusetts, in 1899 and became friends with many writers. One of these writers was Henry James, to whom she dedicated several books. Her first book, a collection of short stories titled *The Greater Inclination*, was published in 1899. She followed that with several novels, including *The Valley of Decision* in 1902 and *The House of Mirth* in 1905.

In 1907, Edith Wharton moved to France. In 1910, her collection of supernatural stories, *Tales of Men and Ghosts*, was published, and her most famous work, *Ethan Frome*, was published in 1911. In 1912, she divorced her husband, who was hospitalized due to a nervous breakdown.

During World War I, Wharton remained in Paris, where her relief work won her a Cross of the Legion of Honor. She described her wartime experiences in the novel *Fighting France*, published in 1915. In 1920, her novel *The Age of Innocence* won a Pulitzer Prize. She continued to write short stories and novels over the next twenty years. In 1934, her autobiography, *A Backward Glance*, appeared. Wharton was the first woman to be awarded an honorary degree from Yale University. She died in France on August 11, 1937. If Indiana had a romance with her during World War I, it would have been even more astonishing than his relationship with Mata Hari. In 1917, Indy would have been seventeen or eighteen at the oldest, while Edith Wharton would have been fifty-five!

As World War I rages on, Indy goes to Instanbul on his next mission, where he uncovers a Turkish scheme involving the murder of French spies.

Originally called Byzantium and then Constantinople, Istanbul is the major city in Turkey today. It is huge and consists of twenty-five districts that lie on both the Asian and European sides of the Bosphorous strait.[36] Located in the northwest Marmara region, Istanbul is 594 square miles in size, and the encompassing Province of Istanbul is 2,402 square miles.

Experts believe that the Asian part of Istanbul was inhabited as far back as 3000 BC. When King Byzas came to the area with colonists in the seventh century to live on the Bosphorous strait, the city became known as Byzantium.

In 324 AD, the Roman emperor Constantine the Great conquered Byzantium. The city became part of the Roman Empire, and by 330 AD, it was called New Rome. It soon was rechristened with the name Constantinople.[37]

The western Roman Empire fell to barbarians in 432 AD, and the eastern part of the empire became known as the Byzantine Empire, with its capital being Constantinople. In 532, during the reign of Justinian I, riots destroyed the city, but it was rebuilt. In fact, it was in constant upheaval, while during the course of the next eight centuries, Persians, Arabs, and Crusaders came in a succession to claim the city as its own.

Then in 1453, Sultan Mehmet II led the Ottoman Turks into the city, and they conquered it, renaming it Istanbul. It became the capital of the Ottoman Empire, and the Turks ruled Istanbul until World War I, when the Allies seized it.

When the Republic of Turkey was created in 1923, Ankara was established as its capital, but Istanbul remained a city. Estimates place Istanbul's current population to be more than thirteen million people.

Dracula and the Eye of the Peacock

In Transylvania, Indy fights vampires. Specifically, he fights the infamous Vlad the Impaler and his horrific army of the living dead. If Indy is successful, he will save humanity.

Popular legend holds that Count Dracula was an ancient vampire and Transylvanian nobleman, who was originally the fifteenth-century Wallachian prince Vlad III. Known as the Impaler, he was a descendant of Attila the Hun. He lived in a decaying castle in the Carpathian Mountains near the Borgo Pass. Though he could be charming and witty, the count was in reality pure evil. According to legend, he possessed the ability to change into a wolf or a bat, had some control of the weather, and was superhumanly strong. Sunlight hurt him, garlic repelled him, and he couldn't stand the sight of a crucifix. He cast no reflection in a mirror and had no shadow. Blood was his only sustenance.

According to the novel by Bram Stoker, in 1897 Dracula moved to London, the capital of the world's greatest empire. There, he infected the wealthy with vampirism until the Dutch vampire hunter Dr. Van Helsing stopped him. Although Dracula fled back to Transylvania, Van Helsing followed him, intent on destroying the evil monster. It seemed that Dracula was indeed destroyed, once and for all, but given that young Indiana Jones encountered him, apparently the vampire rose again.

When a dying man utters, "The eye of the peacock," Indiana and Remy travel from London to Alexandria and even to the South Seas to find Alexander the Great's treasure. This adventure, while thrilling, finally relieves young Indy of his constant fight to save the world during the war.

Alexander the Great, who lived from 356 BC to 323 BC, is one of the most famous figures in history. As the Greek king of Macedonia, Alexander eventually ruled almost all of the known world.

Alexander's father, Philip II, had conquered the many city-states of Greece, but after his death, the southern cities revolted, forcing Alexander to subjugate them again. Afterward, Alexander triumphed over the Persian Empire, which included most of the Middle East. He then conquered all of the kingdoms extending to India. At the time of his death he was planning to vanquish the barbarian kingdoms of Europe.

During his conquests, Alexander came across the legendary Gordian knot in the city of Gordium. The knot was a tangled mass of rope, and a prophecy stated that whoever untangled the knot would rule Asia. Others had tried to untie the knot but had failed.

After looking at the knot, Alexander drew his sword and hacked the rope to pieces. Thus was the Gordian knot untied.

Alexander integrated soldiers from other countries into his army. He also encouraged his soldiers to marry natives from foreign countries. His conquests spread Greek culture and learning throughout much of the known world. After twelve years of constant war, Alexander died at the age of thirty-three.

The Eye of the Peacock was a rare diamond that supposedly once belonged to Alexander the Great. Indiana Jones's friend Remy found on a battlefield a map that marked the location of the diamond. But Indiana decided not to seek the jewel and instead let Remy hunt for it.

In *Temple of Doom*, in 1935 the Chinese gangster Lao Che obtained the famous diamond. Che hired Indiana and Wu Han to find the remains of fabled Chinese emperor Nurhachi, and the duo drew up a false agreement with Che to exchange the remains for the Eye of the Peacock.

Unfortunately, Lao Che planned a double-cross. Wu Han was killed, but Indiana Jones escaped from Lao Che's clutches with his assistant, Shorty, and nightclub singer Willie Scott. What happened to the Eye of the Peacock was never revealed.

In the South Seas, Indy must fight Chinese pirates, only to find himself marooned on a remote desert island. Headhunters capture him, thrilled to think that they might have acquired a good meal and a shrunken head trinket. But along comes anthropologist Bronislaw Malinowski, who saves Indy.

Polish anthropologist Bronislaw Kasper Malinowski was born in 1884. He contributed significantly to ethnographic fieldwork and to the body of knowledge about Melanesia.

As a child, Malinowski was constantly suffering with bad health, but he was a gifted student who earned a doctorate in mathematics and science from Jagiellonian University. Following his studies at the university, Malinowski worked with Wilhelm Wundt on folk psychology at Leipzig University, where he became especially fascinated by anthropology.

In 1914, Malinowski was in Mailu and the Trobriand Islands, performing fieldwork for which he became quite well known. By the time World War I erupted, he was a Polish citizen living in British territory, where he was forced to remain.[38]

At first, Malinowski avoided the islanders because he considered them savages. Then, as time passed, Malinowski learned their language, and out of sheer boredom and loneliness he became fully active in their society. He did major fieldwork on Kula and devised his anthropological method theories about making detailed observations of people. This became known as participant observation.[39]

Without the outbreak of war and isolation, Malinowski most likely would never have created the ideas that became fundamental aspects modern anthropology.

In 1922, Malinowski earned his doctorate in the science of anthropology and published his book *Argonauts of the Western Pacific*, which established him as an authority. He would later, over the course of his career, turn the London School of Economics into a leading institution for studying anthropology.[40]

Malinowski taught some college courses in the United States, and given the nightmare of World War II, he remained in the States and began teaching at Yale University. He died in 1942, while still on the faculty.

Winds of Change: Back in the United States

When World War I ends, Indiana works as a translator in Paris. It is here that he comes into contact with T. E. Lawrence and Prince Faisal of Arabia. Indiana becomes frustrated and sad about the treaty negotiations and returns to the United States. But back in Princeton, New Jersey, he is equally disgusted by the prejudices faced by his boyhood friend Paul Robeson, who is now an actor.

In real life, Indiana's friend Paul Robeson was born and raised in Princeton, New Jersey. He attended Rutgers University in New

Brunswick, New Jersey. Robeson earned fifteen varsity letters in sports and was a two-time All-American football player. He graduated as valedictorian of his class.

Robeson earned his law degree from Columbia University in New York. He was hired by a major law firm but quit when a white secretary refused to take dictation from him.

Making a drastic career change, Robeson won laurels as an actor and a singer. He starred on the Broadway and London stages in numerous roles. He was lionized for his performance of Joe in *Show Boat*, whom he played in the London and New York versions of the show, as well as in the first movie adaptation in 1936. Robeson could sing in twenty different languages.

A strong advocate of civil rights and an outspoken voice against lynching, Robeson was a controversial figure in the late 1940s and early 1950s. He died at age seventy-seven in 1976.

Attending the University of Chicago while working in a speakeasy to pay for his tuition, Indy meets jazz great Sidney Bechet. It is Bechet who teaches Indy how to play the blues.

One of the first jazz soloists to be recorded, Sidney Bechet played both the saxophone and the clarinet. He was born in New Orleans, and at first became well known as an extraordinary jazz clarinetist. Eventually, however, he switched to soprano saxophone and became known as the best saxophonist of his time.

Bechet left New Orleans at nineteen and divided his time between Chicago, New York, Brooklyn, London, and Paris for the next thirty years. In 1950 he finally settled in France, where he married Elisabeth Ziegler in 1951. He died on his sixty-second birthday.

Indiana next encounters the notorious Al Capone. Indy must turn to the future "Untouchable" Eliot Ness for help solving a murder and barely escapes death himself.

Indy's gangster foe Alphonse Gabriel Capone, also known as Al Capone and "Scarface," was once labeled Public Enemy Number 1. One of the most scandalous crime lords of prohibition, Capone was the man most likely responsible for the infamous "St. Valentine's Day Massacre." Like many criminal leaders, Capone used gang members to commit his crimes, thus protecting himself from imprisonment. Capone was finally brought to justice by Eliot Ness and "the Untouchables," who made him stand trial for income tax evasion.

In 1931, Capone was charged in federal court with not paying income tax on his illegal earnings. On the advice of his lawyers, he pleaded guilty to charges of tax evasion. His legal team was hoping for a plea bargain. Instead, the judge sentenced him to eleven years in federal prison. Capone served only six and a half years; his sentence was halved for good behavior. But by the time he was released from prison, his criminal empire had crashed to ruin. Suffering from untreated syphilis that he had gotten as a teenager, he died from cardiac arrest in 1947.

Elliot Ness was a crime fighter who became famous for enforcing Prohibition in Chicago. He was the leader of a team of Treasury agents known as the Untouchables because they couldn't be bribed or paid off by criminals.

Ness graduated from the University of Chicago in 1925 with a degree in business law. In 1926, his brother-in-law, who was a member the Bureau of Investigation, persuaded Ness to join the Treasury Department. Ness worked for the Bureau of Prohibition in Chicago.

After the election of Herbert Hoover as president, Andrew Mellon was given the specific task of bringing down Capone's criminal empire. Ness was put in charge of the operation to destroy all of Capone's illegal breweries. In 1931, he helped to bring Scarface to justice.

After the fall of Capone, Ness was promoted to chief investigator of the Bureau of Prohibition for Chicago. When Prohibition ended in 1933, Ness went to work in Cleveland as director of public safety. He died in 1957 at the age of fifty-four, right before publication of his book *The Untouchables*.

While it made a wonderful example of "what if" storytelling, Indiana Jones couldn't have met Eliot Ness as described in the *Young Indiana Jones Chronicles*. According to the story, Indy met Ness at

the University of Chicago, where they were roommates in April–
May 1920. Ness, however, had just turned seventeen in April 1920
and was a senior in high school. Ness didn't attend the University
of Chicago until 1921. It might have made good television, but the
fabled meeting between Indy and Ness could not have happened.

*Indy, thinking his troubles might finally be over, starts living it up in New
York City. He reads poetry with bohemians, attends lavish high-society
parties, and even directs a Broadway musical. He even hangs out with the
Algonquin Round Table. In New York, Indy meets the famous composer
George Gershwin, who aids Indy in his theatrical ambitions.*

The Algonquin Round Table that Indy encountered in Manhattan
was a real group that existed from 1919 through approximately
1929. The group, consisting of actors, writers, and others, met for
lunch every day at the Algonquin Hotel. There was no formal
membership to the round table, so people came and went. Primary
members included:

Franklin Pierce Adams, columnist

Tallulah Bankhead, stage and screen actress

Robert Benchley, humorist and actor

Heywood Broun, sportswriter

Marc Connelly, playwright

Edna Ferber, playwright

Jane Grant, journalist

Ruth Hale, journalist

Beatrice Kaufman, editor and playwright

George S. Kaufman, playwright and director

Harpo Marx, actor, comedian, and musician

Neysa McMein, magazine illustrator

Dorothy Parker, poet and writer

Harold Ross, the *New Yorker* editor

Robert Sherwood, playwright

Donald Ogden Stewart, playwright and screenwriter

Deems Taylor, composer

John Peter Toohey, publicist

Alexander Woollcott, journalist

Others from the theater and publishing worlds often visited the Round Table. Since several of the members were popular newspaper columnists who repeated some of the conversations in their columns, the quips got wide circulation. Most of their comments were nasty enough to earn them a second nickname, "the vicious circle."

A few examples, as quoted on the PBS series *American Masters: the Algonquin Round Table*, include:[41]

George S. Kaufman: Once when asked by a press agent, "How do I get my leading lady's name into your newspaper?" Kaufman replied, "Shoot her."

Robert Sherwood: Reviewing cowboy hero Tom Mix, Sherwood quipped, "They say he rides as if he's part of the horse, but they don't say which part."

George Gershwin is one of the most famous American composers, both for Broadway and for classical concerts. He wrote popular songs and often collaborated with his brother, Ira.[42] Indy met George soon after he wrote his first big hit, "Swanee," which appeared in the Broadway musical *Broadway Brevities of 1920*.

In 1924, George and Ira Gershwin wrote *Lady Be Good*, which included many popular songs such as "The Man I Love." The Gershwin brothers were the toast of Broadway.

In 1926, they followed *Lady Be Good* with *Oh, Kay*. The next year saw another classic, *Funny Face*. *Strike Up the Band* and *Girl Crazy* followed in 1930. *Of Thee I Sing* played Broadway in 1931 and was the first musical comedy ever to win a Pulitzer Prize.

In between composing work for the musical theater, George Gershwin wrote serious music for the piano. *Rhapsody in Blue* debuted in 1924, performed by Paul Whiteman's concert band. It proved to be extremely successful.

While living in Paris, George Gershwin wrote *An American in Paris*, and then in 1935 he wrote the enormously successful *Porgy and Bess.*[43] Early in 1937, Gershwin's health nosedived, and a doctor determined that he had a brain tumor. He died that year on July 11.

Indy in Hollywood

Indiana moves from Broadway to Hollywood and encounters megalomaniacal director Erich von Stroheim, who argues constantly with Indy about the budget of Stroheim's film Foolish Wives.

Erich von Stroheim was born in Vienna, Austria, in 1885. As a young man he worked in his father's straw hat factory. In 1909, he emigrated to the United States and in 1914 started working for D. W. Griffith as an actor and an assistant director. After the United States entered World War I, von Stroheim found his niche playing arrogant German army officers. He was nicknamed "the man you love to hate."

Von Stroheim wanted to be a director, and in 1919, he directed *Blind Husbands*, which he wrote. He also played the lead character. The film was a commercial success. Unfortunately, Von Stroheim was a perfectionist who liked to produce long films. His film version of Frank Norris's novel *McTeague*, which von Stroheim renamed *Greed*, ran more than nine hours before it was cut by the studio to 140 minutes. Von Stroheim's downfall was the 1929 film *Queen Kelly*, which ran way over budget and was never completed.

No longer able to find work as a director, von Stroheim once again turned to acting. His most immortal role was his last, playing Max von Mayerling in the 1950 movie *Sunset Boulevard*. Von Stroheim spent the last years of his life in France. He died on May 12, 1957.

Foolish Wives was a 1922 film directed by Erich von Stroheim. The reason Indy fought with von Stroheim was that the film's budget came in at more than $1 million. It was the most expensive movie made in Hollywood up to that time. The film starred von Stroheim and Mae Busch, with the story by von Stroheim. The main reason the film cost so much was its elaborate set, a re-creation of Monte Carlo built on the Universal Pictures lot in

Hollywood. The film dealt with a Russian count who seduced foolish women in Monte Carlo for their money and how he was finally murdered by his jilted housekeeper.

The complete movie ran just about eight hours. Von Stroheim cut it down to three and half hours for its premier, but the studio hacked it to a hundred minutes for national release. The film, a bleak, depressing look at human folly, did not earn back its costs.

Although it seems that Indiana can survive anything, he is almost destroyed by the Hollywood movie industry. Never one for giving up a fight, Indy takes one more crack at the business, working with legendary director John Ford. Wyatt Earp and Ford show Indy the real magic of the movies, and in the end, Indy must save the film from disaster when an actor accidentally is killed.

John Ford was a gifted and famous director of early American film, who started his career as an actor in 1914. Ford employed many of his favorite actors repeatedly in his films. For example, John Wayne was in twenty Ford movies. Other standard Ford actors included Ken Curtis, Victor McLaglen, John Carradine, Ben Johnson, and Chill Wills.

Ford made his reputation with his Westerns, but he directed all types of films. He was the only director to win four Academy Awards for Best Director, for *The Informer*, *The Grapes of Wrath*, *How Green Was My Valley*, and *The Quiet Man*. Ford died in Palm Desert, California, at age seventy-nine, from stomach cancer.

Wyatt Berry Stapp Earp was born in Monmouth, Illinois, and grew up on a farm in Iowa. In 1864, he moved with his parents to California. After working as a stagecoach driver and a buffalo hunter, he served as deputy marshal in Wichita, Kansas, and Dodge City, Kansas, where he became friends with Bat Masterson and Doc Holliday. Over the years, he developed a reputation as a lawman and a gambler.

In Tombstone, Arizona, Earp acquired the gambling concession at the Oriental Saloon. In 1881, a feud with the Clanton gang ended with the famous Gunfight at the OK Corral. Three of the Clanton gang were killed. The events of the gunfight became obscured in the many retellings until the gunfight turned into a Wild West legend.

Wyatt Earp spent his final years living off his real estate and mining investments. He and his wife befriended a number of early Hollywood actors, including John Wayne. Earp died at the age of eighty on January 13, 1929.

It seems appropriate that a modern-day legend like Indiana Jones, in his last adventure as a young man, should meet a legendary historical figure like Wyatt Earp.

PART 5

INDIANA JONES

and the
Kingdom of the Crystal Skull

XOX XOX XOX

The Secrets of the Real Crystal Skull Revealed

The fourth Indiana Jones movie, *Indiana Jones and the Kingdom of the Crystal Skull*, finds our hero engaged in a search for an ancient relic of somewhat dubious power and origins: a crystal skull. According to certain New Age philosophers, ancient crystal skulls serve as the focal point of radiant psychic energy and improve the life of the owner through handling and being spoken to, as you would a pet. Other New Age mystics claim that crystal skulls work like crystal balls and can be used to tell the future. In simple terms, a crystal skull is made from quartz crystal and is fashioned to look like a human skull.[1]

The most notorious of all crystal skulls was found by Anna Le Guillon Mitchell-Hedges in 1926. The adopted daughter of famous archeologist F. A. "Mike" Mitchell-Hedges, Anna wrote to author Richard Garvin that she had unearthed the crystal skull from beneath a temple in Lubaantun, British Honduras. According to Garvin in his book *The Crystal Skull*, Anna said that a Mayan high priest used the skull to "will death" to his enemies.[2] The crystal skull became known as "the Skull of Doom" because of its supposed psychic properties and the bad luck experienced by people who touched it.

The actual skull was made from pure crystalline quartz, consisting of a single left-handed growing crystal. It measured five inches high, seven inches long, and five inches wide, and it weighed approximately eleven pounds. Its details were nearly perfect.

In 1970, art restorer Frank Dorland examined the skull at the Hewlett-Packard laboratory in California. No lab report was ever issued, but Dorland made a number of claims based on information supposedly supplied to him by technicians who studied the skull. According to Dorland, scientific tests showed that the skull was carved into rough form by diamond chisels. He also claimed that no metal tools were used in shaping the relic, and that the creation of the skull, including the formation of its shape as well as its high polish, occurred over a period of three hundred years. All of which, Dorland declared, proved that the skull was the product of technology more advanced than anything existing at the time.

Dorland also stated that the skull came from the ancient kingdom of Atlantis, and, further, that during the Crusades, it was part of the vast treasure owned by the Knights Templar.

When F. A. Mitchell-Hedges's autobiography, *Danger My Ally*, was published in 1954, he wrote that the crystal skull was over three thousand years old. However, the archaeologist never stated where the skull was found or who found it. He also wrote that when the high priest of the Maya used the skull to will people to die, they died, although again he offered no evidence to back up his statement.

Recently, scholars working at the British Museum discovered records indicating that the crystal skull once belonged to London art and antique dealer Sydney Burney, who sold it to Frederick Mitchell-Hedges in 1944. Thus, the entire episode describing Anna's finding the skull in the ruins of an ancient temple was proven false, but she refused to change her story.

Since the discovery of the crystal skull, many other very similar skulls have been found throughout the world, such as in Peru, Mexico, and the Honduras. New Age devotees claim that all of the skulls originated in Atlantis, but scientists have shown that the skulls probably were made in the nineteenth century in Germany.

Further research determined that several of the skulls owned by European museums were bought from a French antique dealer, Eugne Boban, whose business was located in Mexico City from 1860 to 1880. An investigation carried out by the Smithsonian in 1992 on yet another crystal skull, supposedly of Aztec origin, concluded that it was carved in the last 150 years. A paper trail of bills proved the skull originated with Boban. Researchers concluded that Boban bought his crystal skulls from German craftsmen and then sold them in Mexico City as ancient artifacts.

Despite all the evidence that crystal skulls are not remnants of Atlantis and have no mysterious powers or hold any mystical secrets, there are people who still believe that the skulls are powerful artifacts of an ancient civilization. This proves once again that if you try hard enough, you can fool some of the people all of the time.

Notes

Part 1: Indiana Jones and the Raiders of the Lost Ark

1. See http://news.nationalgeographic.com/news/2003/04/0418_030418 _bibleartifact.html.
2. Michael Parfit, "Hunt for the First Americans," *National Geographic*, December 2000: 40.
3. See http://209.85.165.104/search?q=cache:PatrNuFpKBUJ:graphics.cs .brown.edu/research/sciviz/archaeology/archave/Scientific_Archaeology .pdf+archaeology+is+not+an+exact+science&hl=en&ct=clnk&cd=13&gl=us &client=safari; Brown University's Computer Graphics Group, http:// graphics.cs.brown.edu/.
4. See http://ccat.sas.upenn.edu/bmcr/2005/2005–05–05.html.
5. See *Interaction* newsletter, fall 2006, http://multi.stanford.edu/interaction/ 1106/arch.html.
6. Hiram Bingahm, *Lost City of the Incas*, www.senate.gov/reference/reference _item/LostCity.htm, accessed January 8, 2007.
7. See www.machupicchuexplorer.com.
8. See www.nationalgeographic.com/inca/machu_picchu.html.
9. Institute for Biblical and Scientific Studies, "Biblical Archaeology: Where is the Ark of the Covenant?" www.bibleandscience.com/archaeology/ark.htm, accessed January 8, 2007.
10. Base Institute, "Search for the Ark of the Covenant," www.baseinstitute.org/ features/arkofcovenant.htm, accessed January 8, 2007.
11. John Roach, "Fear of Snakes, Spiders Rooted in Evolution, Study Finds," *National Geographic News*, October 4, 2001, http://news.nationalgeographic .com/news/2001/10/1004_snakefears.html.
12. See www.feldgrau.com/aircraft.html.

Part 2: Indiana Jones and the Temple of Doom

1. "Monkey Brains on the Menu" by Richard C. Paddock, staff writer, *L.A. Times*, www.mongabay.com/external/monkey_brains.htm, accessed

September 24, 2007. The original article is in the *L.A. Times* archives, www
.latimes.com/news/printedition/la-fg-oddeats25feb25001450,1,4363399
.story.

2. *Indian Insect Pests*, published by the superintendent of government printing,
India, Calcutta, by H. Maxwell-Lefroy, M.A., F.E.S., F.Z.S., Imperial Ento-
mologist, 1906, http://ia311519.us.archive.org/3/items/indianinsectpest00
maxwrich/indianinsectpest00maxwrich_djvu.txt.

3. See the extensive chapter on this subject in Lois H. Gresh, *Exploring Phillip
Pullman's His Dark Materials: Dust, Angels, Souls, and Weird Science* (New
York: St. Martin's Press, 2008), chap. 8. This book was written several years
ago but at the time of this writing has not yet been published.

Part 3: Indiana Jones and the Last Crusade

1. Vestil S. Harrison, *Utah History Encyclopedia*, www.media.utah/UHE/f/
FREMONT%2CTHE.html.

2. "Piltdown Man," *Wikepedia*, http://en.wikipedia.org/wiki/Piltdown_Man,
accessed February 22, 2008.

3. Karl Beese, "Giovanni Battista Belzoni," Minnesota State University,
Mankato, 1999, www.mnsu.edu/emuseum/information/biography/abcde/
belzoni_giovanni.html.

4. "The Great Belzoni (1778–1823)," www.belzoni.com/giovanni.htm,
accessed February 22, 2008.

5. "Belzoni—Tomb Raider and Archaeologist: Part 3," www.bbc.co.uk/dna/
h2g2/A22548477, accessed February 22, 2008.

6. "Francisco Vásquez de Coronado," *Wikipedia*, www.en.wikipedia.org/wiki/
Francisco_V%C3%A1squez_de_Coronado, accessed February 22, 2008.

7. Ibid.

9. Ibid.

9. Lavahn Hoh, *The Circus in America 1793–1940*, Institute for Advanced Technology in the Humanities, University of Virginia, 2004, www.circusinamerica
.org/public/people/public_show/71.

10. "Louis Leakey," *Wikipedia*, http://en.wikipedia.org/wiki/Louis_Leakey#
His_father.27s_example, accessed February 22, 2008.

11. "Bundesfuhrer Kuhn," AmericanHeritage.com, www.americanheritage
.com/articles/magazine/ah/1995/5/1995_5_102.shtml, accessed February
22, 2008.

12. Ibid.

13. Ibid.

14. Ibid.

15. Ibid.

16. Interview with Henry Ford, *New York World*, February 1921.

17. "Fallen Hero: Charles Lindbergh in the 1940s," PBS online, www.pbs.org/wgbh/amex/lindbergh/sfeature/fallen.html, accessed February 22, 2008.

18. Ibid.

19. "How Bush's Grandfather Helped Hitler's Rise to Power," www.guardian.co.uk/world/2004/sep/25/usa.secondworldwar, accessed February 22, 2008.

20. Seymour Hersh, *The Dark Side of Camelot* (Boston: Little, Brown, 1997), p. 63.

21. Ibid., p. 64.

22. Laurence Leamer, *The Kennedy Men: 1901–1963: The Laws of the Father* (New York: Harper, 2002), p. 134.

23. Ibid.

24. "The Holy Grail," *New Catholic Encyclopedia*, www.newadvent.org/cathen/06719a.htm, accessed February 22, 2008.

25. Ibid.

26. Ibid.

27. Ibid.

28. "The Holy Grail," *Wikipedia*, http://en.wikipedia.org/wiki/Grail, accessed July 7, 2007.

29. Ibid.

30. Ibid.

31. Ibid.

32. Ibid.

33. "The First Crusade," *Wikipedia*, http://en.wikipedia.org/wiki/First_Crusade, accessed July 7, 2007.

34. Ibid.

35. Ibid.

36. Ibid.

37. "Catacombs," *Wikipedia*, http://en.wikipedia.org/wiki/Catacombs, accessed February 22, 2008.

38. Robert Hendrickson, *More Cunning than Man: A Social History of Rats and Men* (New York: Stein & Day, 1983), p. 5.

39. Ibid.

40. Brorsen, Carol, "Real Life Rat Tales," *Austin Chronicle*, January 21, 2000, www.austinchronicle.com/gyrobase/Issue/story?oid=oid%3A75571.

41. Ibid.

42. Ibid.

43. Peter Rosen, *The Catholic Church and Secret Societies* (Milwaukee: Houtkamp & Cannon, 1902), p. 2.

44. Ibid., p. 3.

45. "The Triumph of Hitler," www.historyplace.com/worldwar2/triumph/tr-bookburn.htm, accessed February 22, 2008.

46. "Helen Keller's Response to Nazi Book Burnings," *Helen Keller's Museum for Kids Online*, American Foundation for the Blind, 2008, www.afb.org/braillebug/hkholocaust.asp.

47. "Zeppelins," *Wikipedia*, http://en.wikipedia.org/wiki/Zeppelin, accessed July 7, 2007.

48. Ibid.

49. Ibid.

50. Ibid.

51. Ibid.

52. Ibid.

53. Ibid.

54. Paul Crawford, "Crusading Vows and Privileges," *ORB Online Encyclopedia*, www.the-orb.net/encyclop/religion/crusades/vows.html, accessed July 7, 2007.

55. Ibid.

56. Karin Laub, "Bible Scholar: Last Supper Chalice Likely a Simple Clay Cup," Associated Press, April 11, 1998, www.uhl.ac/cup.html.

57. Ibid.

Part 4: Young Indiana Jones

1. "Montmarte," *Wikipedia*, http://llen.wikipedia.org/wiki/Montmarte, accessed July 1, 2007.

2. Ibid.

3. Ibid.

4. "About Rockwell—the Official Site of Site of Norman Rockwell, www.normanrockwell.com/about/biography.htm, accessed June 5, 2007.

5. "Norman Rockwell," *Wikipedia*, http://en.wikipedia.org/wiki/Norman_Rockwell, accessed June 5, 2007.

6. "Georges Braque," *Wikipedia*, http://en.wikipedia.org/wiki/Braque, accessed June 1, 2007.

7. Tom Rosenthal, "When Pablo Met Henri," *Independent*, October 30, 2005, http://findarticles.com/p/articles/mi_qn4159/is_20051030/ai_n15816646.

8. "Frederick Selous," www.answers.com/topic/frederick-selous, accessed February 21, 2008.

9. "Sigmund Freud, *Wikipedia*, http://en.wikipedia.org/wiki/Sigmund_Freud, accessed May 15, 2007.

10. Chris Roberts, "Pancho Villa, the Marketing Whiz," *Milwaukee Journal Sentinel*, September 3, 2003.

11. Quotation from the Nobel Prize in Literature, 1923.

12. John Simkin, "Zeppelin Raids," *Spartacus Educational*, www.spartacus .schoolnet.co.uk/FWWzeppelinraids.htm, accessed May 1, 2007.

13. Ibid.

14. Ibid.

15. "Battle of the Somme," *Wikipedia*, http://en.wikipedia.org/wiki/Battle_of_ the_Somme_(1916), accessed February 22, 2008.

16. "Trench Warfare," http://encarta.msn.com/encyclopedia_761585472/ Trench_Warfare.html, accessed February 22, 2008.

17. "World War I," http://encarta.msn.com/encyclopedia_761569981_5/ World_War_I.html#p83, accessed February 22, 2008.

18. "1918, Hell on Earth," *Trench Warfare*, www.awm.gov.au/1918/trenchwarfare/ index.htm, cited May 1, 2007.

19. "Poison Gas and World War I," *Learning Site*, www.historylearningsite .co.uk/poison_gas_and_world_war_one.htm, accessed May 25, 2007.

20. Ibid.

21. Ibid.

22. "Jan Smuts," *Wikipedia*, http://en.wikipedia.org/wiki/Jan_Smuts, accessed May 21, 2007.

23. Time Life Books, *Knights of the Air* (Time Life Books Aviation Series, 1996), p. 120.

24. "Lafayette Escadrille," *Wikipedia*, http://en.wikipedia.org/wiki/Lafayette_ Escadrille, accessed May 3, 2007.

25. "Manfred von Richthofen," *Wikipedia*, http://en.wikipedia.org/wiki/ Manfred_von_Richthofen, accessed April 29, 2007.

26. "Who Is the Red Baron?"_essortment, www.essortment.com/whoredbaron _rkzx.htm, accessed April 29, 2007.

27. "Manfred von Richthofen," *Wikipedia*, http://en.wikipedia.org/wiki/ Manfred_von_Richthofen, accessed April 29, 2007.

28. "Anthony Fokker," *Wikipedia*, http://en.wikipedia.org/wiki/Anthony_ Fokker, accessed April 30, 2007.

29. Ibid.

30. Peter Drucker, *Managing in the Next Society* (New York: St. Martins, 2003), p. 33.

31. "Boshevik," *Wikipedia*, http://en.wikipedia.org/wiki/Bolshevik accessed May 5, 2007.

32. "Beersheba—the Only Great Mounted Infantry Charge in History," www.diggerhistory.info/pages-conflicts-periods/ww1/lt-horse/beersheba .htm, accessed May 20, 2007.

33. Steve Paul, "Bigger Than Life: Ernest Hemingway at 100," *Kansas City Star*, www.kcstar.com/hemingway.

34. "French Foreign Legion," *Wikipedia*, http://en.wikipedia.org/wiki/French_ foreign_legion, accessed May 22, 2007.

35. Ibid.

36. "Istanbul," *Wikipedia*, http://en.wikipedia.org/wiki/Istanbul, accessed June 5, 2007.

37. "Constantine the Great," *All about Turkey*, www.allaboutturkey.com/ konstantin.htm, accessed February 22, 2008.

38. "Bronislaw Malinowski," *Wikipedia*, http://en.wikipedia.org/wiki/Bronislaw_ Malinowski, accessed May 29, 2007.

39. Bronson Malinowski, *Argonauts of the Western Pacific* (New York: Dutton, 1961), p. 25.

40. "Bronislaw Malinowski," *Wikipedia*, http://en.wikipedia.org/wiki/Bronislaw_ Malinowski accessed May 29, 2007.

41. "The Algonquin Round Table," PBS online, www.pbs.org/wnet/ americanmasters/database/algonquin_round_table.html, accessed February 22, 2008.

42. "George Gershwin," *Wikipedia*, http://en.wikipedia.org/wiki/George_ Gershwin, accessed June 15, 2007.

43. Ibid.

Part 5: Indiana Jones and the Kingdom of the Crystal Skull

1. "Crystal Skull," *Wikipedia*, http://en.wikipedia.org/wiki/Crystal_skull, accessed February 22, 2008.

2. Ibid.

Bibliography

"Algonquin Round Table." *Art and Culture*. www.artandculture.com/cgi-bin/WebObjects/ACLive.woa/wa/movement?id=454, accessed May 25, 2007.

"Alfred Vincent Kidder, Biography." *The Columbia Electronic Encyclopedia*. 6th ed. New York: Columbia University, 2003.

Andrews, Mark. "Egypt: An Overview of the West Bank at Luxor." *Tour Egypt*. InterCity Oz, Inc., 1999–2005. www.touregypt.net/featurestories/westbank.html, accessed June 5, 2007.

"Anthony Fokker." *Wikipedia*. http://en.wikipedia.org/wiki/Anthony_Fokker, accessed June 20, 2007.

"Archaeologists, Biographies." *About.com. Archaeology*. http://archaeology.about.com/od/archaeologistsuv/Archaeologist_Biographies_htm, accessed June 20, 2007.

Aris, Ben, Duncan Campbell. "How Bush's Grandfather Helped Hitler's Rise to Power." *Guardian Unlimited*, September 25, 2004. www.guardian.co.uk/usa/story/0,12271,1312540,00.html, accessed May 27, 2007.

"Ark of the Covenant." *Artifacts, Marshall College*. www.indianajones.com/marshall/artifact/arkofthecovenant/, accessed June 6, 2007.

Asbridge, Thomas. *The First Crusade: A New History*. New York: Oxford, 2004.

Augustine, Keith. "The Case against Immortality." *Skeptic Magazine*, 1997. www.infidels.org/library/modern/keith_augustine/immortality.html, accessed May 30, 2007.

Beese, Karl. "Giovanni Battista Belzoni." Makato: Minnesota State University, 1999. www.mnsu.edu/emuseum/information/biography/abcde/belzoni_iovanni.html, accessed June 1, 2007.

Bellis, Mary. "Ferdinand von Zeppelin." *About.com*. http://inventors.about.com/od/xzstartinventors/ss/Zeppelin.htm, accessed June 5, 2007.

"Belzoni—Tomb Raider and Archaeologist." *Life The Universe Everything*. www.bbc.co.uk/dna/h2g2/A22548477, accessed May 25, 2007.

"Bible-Era Artifacts Highlight Archaeology Controversy." *National Geographic News*. http://news.nationalgeographic.com/news/2003/04/0418_030418_bibleartifact.html, accessed June 5, 2007.

"Bolsheviks." *Spartacus Educational.* www.spartacus.schoolnet.co.uk/RUSbolshe viks.htm, accessed May 28, 2007.

Brorsen, Carol. "Real Life Rat Tales." *Austin Chronicle*, January 21, 2000. www.austinchronicle.com/gyrobase/Issue/story?oid=oid%3A75571, accessed June 5, 2007.

Brown University's Computer Graphics Group. http://graphics.cs.brown.edu/, accessed June 20, 2007.

"Bundesfuhrer Kuhn." *AmericanHeritage.com*, 2006. www.americanheritage.com/ articles/magazine/ah/1995/5/1995_5_102.shtml, accessed June 6, 2007.

"The Burning of Books." *The History Place.* www.historyplace.com/worldwar2/ triumph/tr-bookburn.htm, accessed June 15, 2007.

Carter, D. "Why Mormonism?" 2006. *Mormonism Research Ministry.* http:// www.mrm.org/about/why-mormonism, accessed June 1, 2007.

"Catacombs." *Wikipedia.* http://en.wikipedia.org/wiki/Catacombs, accessed June 18, 2007.

"Charles de Gaulle." *BBC Co.*, UK. www.bbc.co.uk/history/historic_figures/ gaulle_charles_de.shtm, accessed June 10, 2007.

"Charles de Gaulle." *Charles de Gaulle.* http://econ161.berkeley.edu/TCEH/ charlesdegaulle.html, accessed June 10, 2007.

"Circus History." *CircusWeb History of the Circus*, Graphics 2000 and Intelligent Software Associates, Inc. 1996. www.circusweb.com/cwhistory.html, accessed May 20, 2007.

Cirlot, Juan-Eduardo. *Picasso: Birth of a Genius.* New York: Praeger, 1972.

"Conquistadors." *PBS.org.* Text derived from "Conquistadors" by Michael Wood. University of California Press, Spring 2001. www.pbs.org/conquista dors/, accessed May 17, 2007.

Crystal, Ellie. "King Tutankhamen's Tomb." *Ellie Crystal's Metaphysical and Science Site*, 1995–2007. www.crystalinks.com/tutstomb.html, accessed May 21, 2007.

"Crystal Skull." Wikipedia. http://en.wikipedia.org/wiki/Crystal_skull, accessed December 5, 2007.

"Dietrich Eckhart." *Oxford University Press. Answers.com.* www.answers.com/ topic/dietrich-eckart?cat=entertainment, accessed June 5, 2007.

Drucker, Peter. *Managing in the Next Society.* New York: St. Martins, 2003.

Du Bois, and Shirley Graham. *Paul Robeson, Citizen of the World.* Westport: Julian Messner, 1971.

Duffy, Michael. "Battles: The Battle of Somme, 1916." *First World War.com.* www.firstworldwar.com/battles/somme.htm, accessed June 14, 2007.

Duffy, Michael. "Who's Who: John Pershing." *First World War.com.* www.first worldwar.com/bio/pershing.htm, accessed June 14, 2007.

Emick, Jennifer. "Grail Cross." *About.com*, 2007, About, Inc. http://altreligion .about.com/library/glossary/symbols/bldefsgrailcross.htm, accessed June 20, 2007.

"Erich Von Stroheim." *Sunset Boulevard: The Movie Web Site*. www.geocities.-com/Hollywood/theater/6980/von.htm, accessed June 16, 2007.

"Ernest Hemingway's 'The Killers.'" *"The Killers," An Annotated Bibliography*. http://caxton.stockton.edu/FEL08LITT2143/stories/storyReader$8, accessed June 20, 2007.

"Fallen Hero: Charles Lindbergh in the 1940s." *The American Experience. PBS Online*. www.pbs.org/wgbh/amex/lindbergh/sfeature/fallen.html, accessed May 17, 2007.

"The First Crusade." *Wikipedia*. http://en.wikipedia.org/wiki/First_Crusade, accessed May 25, 2007.

"Frederick Selous." *Answers.com*. www.answers.com/topic/frederick-selous, accessed May 28, 2007.

Gannon, Michael. *Black May: The Epic Story of the Allies' Defeat of the German U-Boats in May 1943*. New York: HarperCollins, 1998.

Garvin, Richard. *The Crystal Skull: The Story of the Mystery, Myth and Magic of the Mitchell-Hedges Crystal Skull Discovered in a Lost Mayan City during a Search for Atlantis*. New York: Doubleday, 1973.

"German American Bund." *Wikipedia*. http://en.wikipedia.org/wiki/German _American_Bund.

"Grave Robbing." *Wikipedia*. http://en.wikipedia.org/wiki/Grave_robbing, accessed June 6, 2007.

"The Great Belzoni (1778–1823)." *Belzoni Online*. http://www.belzoni.com/gio vanni.htm, accessed May 25, 2007.

Gresh, Lois H. *Exploring Phillip Pullman's His Dark Materials: An Unauthorized Adventure through the Golden Compass, the Subtle Knife, and the Amber Spyglass*. New York: St. Martin's Press, 2008.

Grossman, Daniel J. "DJ's Zeppelin Page: Airships." *Airships*, www.airships.net/, accessed June 10, 2007.

Harrison, Vestil S. "Centerville." *Utah History Encyclopedia*. www.media.utah .edu/UHE/c/CENTERVILLE.html, accessed May 20, 2007.

Harter, Richard. "Piltdown Man." *Richard Harter's World*, October 30, 2006. http://home.tiac.net/~cri_a/piltdown/piltdown.html, accessed June 2, 2007.

Helen Keller's Response to Nazi Book Burnings. *Helen Keller's Museum for Kids Online*, copyright 2007 by American Foundation for the Blind. www.afb.org/braillebug/hkholocaust.asp, accessed May 25, 2007.

"Hell on Earth: Trench Warfare." *Australian War Memorial*. http://www.awm .gov.au/1918/trenchwarfare/, accessed June 15, 2007.

Hendrickson, Robert. *More Cunning Than Man: A Social History of Rats and Men*. New York: Stein & Day Books, 1983.

Hersh, Seymour. *The Dark Side of Camelot*. New York: Back Bay Books, 1998.

"Hiram Bingham." *Machu Picchu Explorer*. www.machupicchuexplorer.com/, accessed May 27, 2007.

Hoh, Lavahn. *The Circus in America 1793–1940*. University of Virginia, The Institute for Advanced Technology in the Humanities, 2004. www.circusinamerica .org/public/people/public_show/71, accessed June 8, 2007.

"Holy Blood, Holy Grail." *Wikipedia*. http://en.wikipedia.org/wiki/Holy _Blood_Holy_Grail, accessed May 26, 2007.

"The Holy Grail." *Artifacts. Marshall College*, 2007, Lucasfilm Ltd. www.indiana jones.com/marshall/artifact/holygrail/, accessed June 5, 2007.

"Indiana Jones and the Last Crusade." *Answers.com*. www.answers.com/topic/ indiana-jones-and-the-last-crusade?cat=entertainment, accessed May 20, 2007.

"Indiana Jones and the Last Crusade." *IMDb*. www.imdb.com/title/tt0097576/ goofs, accessed May 21, 2007.

Indian Insect Pests, published by the Superintendent of Government Printing, India, Calcutta, by H. Maxwell-Lefroy, MA, FES, FZS, Imperial Entomolo gist, 1906. *The Library of the University of California*, http://ia311519.us.archive .org/3/items/indianinsectpest00maxwrich/indianinsectpest00maxwrich_djvu .txt, accessed May 25, 2007.

"In Search of the Ark of the Covenant." *Base Institute Web Site*, www.baseinstitute .org/features/arkofcovenant.htm, accessed June 20, 2007.

"Interview with Henry Ford." *New York World*, February, 1921.

"J." *Wikipedia*. http://en.wikipedia.org/wiki/J, accessed June 1, 2007.

"Jan Smuts." *Wikipedia*. http://en.wikipedia.org/wiki/Jan_Smuts, accessed June 6, 2007.

"Jim Thorpe, World's Greatest Athlete." *CMG Worldwide*, August 2, 2005. www.cmgworldwide.com/sports/thorpe/, accessed June 2, 2007.

Kahler, Wendy. "History of the Circus." *essortment*, 2002 Pagewise. http://ok .essortment.com/circus_rnws.htm, accessed May 21, 2007.

Katz, Bob. "A Most Famous Failing." *Desert USA*. www.desertusa.com/mag98/ sep/papr/coronado.html, accessed May 29, 2007.

"King Tutankhamen's Tomb." *Ellie Crystal's Metaphysical and Science Website*, www.crystalinks.com/tutstomb.html, accessed May 22, 2007.

"Knowledge of archaeology. " *Multidisciplinary Teaching and Research, Stanford University*. http://multi.stanford.edu/interaction/1106/arch.html; Fall 2006 *Interaction* newsletter, accessed May 27, 2007.

"The Last Great Mounted Charge." *Defending Victoria Web Site*. http://users .netconnect.com.au/~ianmac/bersheba.html, accessed June 5, 2007.

Laub, Karin. "Bible Scholar: Last Supper Chalice Likely a Simple Clay Cup." Associated Press, April 11, 1998. www.uhl.ac/cup.html accessed June 6, 2007.

Leamer, Laurence. *The Kennedy Men: 1901–1963*. New York: Harper, 2002.

Leigh, Richard, and Henry Lincoln. *Holy Blood, Holy Grail*. New York: Dell, 1983.

"Light Horse." *Wikipedia*. http://en.wikipedia.org/wiki/Light_horse, accessed June 5, 2007.

"The Lost Ark." *Bryn Mawr Classical Review*. http://ccat.sas.upenn.edu/bmcr/ 2005/2005–05–05.html, accessed June 6, 2007.

"Lost City of the Incas." *U.S. Senate Reference*. www.senate.gov/reference/refer ence_item/LostCity.htm, accessed June 8, 2007.

"Luftwaffe-Aircraft." *GermanWarMarchine.com*. www.feldgrau.com/aircraft.html, accessed May 29, 2007.

"Machu Picchu Flashback." *Secrets of a Lost World*. www.nationalgeographic .com/inca/machu_picchu.html, accessed May 27, 2007.

Madden, Thomas F. *Crusades, an Illustrated History*. East Lansing: University of Michigan Press, 2005.

Malinowski, Bronson. *Argonauts of the Western Pacific*. New York: Dutton, 1961.

"Manfred von Richthofen." *Wikipedia*. http://en.wikipedia.org/wiki/Manfred _von_Richthofen, accessed May 25, 2007.

Meyer, Ronald Bruce. "Nazi Book-Burning (1933): Religion and Censorship." *Ronald Bruce Meyer.com*. www.ronaldbrucemeyer.com/rants/0510almanac.htm. 1980.

Mitchell-Hedges, F. *Danger My Ally*. Kempton, IL: Adventures Unlimited Press, 1995.

"Mormon Settlement." *Utah History for Kids*. State of Utah, 2007.

Murray, Craig. *Boy Scouts of America Badge History Page*, March 28, 2005. www .sageventure.com/history/index.htm, accessed May 25, 2007.

1912 Olympics—Stockholm. *Fact Monster Sports*. Fact Monster/Information Please® Database 2007, Pearson Education, Inc. www.factmonster.com/ipka/ A0114419.html, accessed May 30, 2007.

Noe, Denise. "All About Mata Hari." *CourtTV Crime Library*. www.crimelibrary .com/spies/mata_hari/, accessed May 30, 2007.

Nowak, Jeff, and Allen B. Ruch. "The Most Fortunate and Unfortunate of Men." *Franz Kafka Biography*. *The Modern Word.com*, January 22 2004. www.themodernword.com/kafka/kafka_biography.html, accessed June 7, 2007.

Orans, Lewis P. *Robert Baden-Powell, Founder of the World Scout Movement, Chief Scout of the World*. New York: Lewis P. Orans, 2004.

"Pancho Villa Biography." *Biography.com*, www.biography.com/search/article .do?id=9518733, accessed June 20, 2007.

Parfit, Michael. "Hunt for the First Americans." *National Geographic*, December, 2000.

Parsons, Marie, "Tomb-Robbery." www.touregypt.net/featurestories/robbery.htm accessed May 21, 2007.

"Paul Robeson." *Wikipedia*. http://en.wikipedia.org/wiki/Paul_Robeson, ac-cessed June 8, 2007.

Paul, Steve. "Bigger Than Life: Ernest Hemingway at 100." *Kansas City Star*. June 27, 1999. www.kcstar.com/hemingway/, accessed June 8, 2007.

"Poison Gas." *History Learning Site.com*. www.historylearningsite.co.uk/poison _gas_and_world_war_one.htm, accessed May 19, 2007.

Porter, Perry. *LDS Church History*. 1830–1904. *LDS Church History.com*. www.ldshistory.net/, accessed June 6, 2007.

Powers, John. "Searching for the Artifacts of Faith: Examining the Power and Mystery Surrounding Easter Relics, from the Holy Grail to Pieces of the Cross to the Controversial Shroud of Turin." *Insight on the News*, April 27, 2004.

Quotation from the Nobel Prize in Literature, 1923.

Reeve, W. Paul. "The First Cars in Two Small Towns." *Utah History to Go*, October 1996. http://historytogo.utah.gov/utah_chapters/statehood_and_the _progressive_era/thefirstcarsintwosmalltowns.html, accessed May 25, 2007.

Rice, Boyd. "The Cross of Lorraine: The Fingerprint of God or the Mark of Cain?" *Discriminate Media*, 2006. *The Grail Research & Esoteric Writings of Boyd Rice*. www.thevesselofgod.com/home.html, accessed May 22, 2007.

Rich, Tracey R. "Passover." *Judaism 101*, 5756–5766 (1995–2005). www.jewfa-q.org/holidaya.htm, accessed June 17, 2007.

Roach, John. "Fear of Snakes, Spiders Rooted in Evolution, Study Finds." *National Geographic News*, October 4, 2001. http://news.nationalgeographic.com/ news/2001/10/1004_snakefears.html, accessed May 29, 2007.

Roberts, Chris. "Pancho Villa, the Marketing Whiz." *Milwaukee Journal Sentinel*, September 3, 2003.

Rosen, Peter. *The Catholic Church and Secret Societies*. Hollendale: Houtkamp & Cannon, 1902.

Rosenthal, Tom. "When Pablo Met Henri." *Independent*, October 30, 2005. http://findarticles.com/p/articles/mi_qn4159/is_20051030/ai_n15816646, ac-cessed June 1, 2007.

Sansal, Burak. "History of Istanbul." *All about Turkey*. www.allaboutturkey.com/ istanbul.htm, accessed May 19, 2007.

Schwichtenberg, Holly. "Flinders Petrie." *Flinders Petrie*, Minnesota State Uni-versity, 2006. www.mnsu.edu/emuseum/information/biography/pqrst/petrie _flinders.html, accessed June 7, 2007.

"Scientific Archaeology vs. the Discovery Channel." *Archaeology, Brown University*. http://209.85.165.104/search?q=cache:PatrNuFpKBUJ:graphics.cs.brown.edu/ research/sciviz/archaeology/archave/Scientific_Archaeology.pdf+archaeology+is +not+an+exact+science&hl=en&ct=clnk&cd=13&gl=us&client=safari, accessed June 1, 2007.

"Scouting in Utah." *Wikipedia*. http://en.wikipedia.org/wiki/Scouting_in_Utah, accessed May 22, 2007.

Sharp, Jay. *The Anasazi: The People of the Mountains, Mesas and Grasslands*. Desert *USA*. www.desertusa.com/ind1/du_peo_ana.html, accessed May 22, 2007.

Shermer, Michael. "Illuminati: The New World Order & Paranoid Conspiracy Theorists." eSkeptic, Skeptics Society, *The Skeptics Dictionary*. http://skepdic .com/illuminati.html, 2006, Michael Shermer & the Skeptics Society, accessed June 7, 2007.

Simkin, John. "Anthony Fokker." *Spartacus Educational*. www.spartacus.school net.co.uk/FWWfokker.htm, accessed May 30, 2007.

"State of Deseret." *Wikipedia*. http://en.wikipedia.org/wiki/State_of_Deseret, accessed June 6, 2007.

"St. Stephen's Cathedral, Vienna." *Wikipedia*. http://en.wikipedia.org/wiki/ St._Stephen%27s_Cathedral%2C_Vienna, accessed May 25, 2007.

"Tell Me More about Adolf Hitler and WWII." *ThreeWorldWars.com*. http:// threeworldwars.com/world-war-2/adolf-hitler.htm, accessed June 8, 2007.

"Thule Society." *Wikipedia*. http://en.wikipedia.org/wiki/Thule_Society, accessed June 17, 2007.

Time Life Books. *Knights of the Air*. New York: Time Life Books Aviation Series, 1996.

"Trench Warfare." *Wikipedia*. http://en.wikipedia.org/wiki/Trench_warfare, accessed June 7, 2007.

Trueman, Chris. "Poison Gas and World War One." *History Learning Site*, www.historylearningsite.co.uk/poison_gas_and_world_war_one.htm, accessed June 1, 2007.

Warren, John. "Howard Carter." *Tour Egypt*, 1996–2007, by InterCity Oz, Inc. www.touregypt.net/featurestories/carter.htm, accessed June 10, 2007.

"Where Is the Ark of the Covenenant?" *Institute for Biblical and Scientific Studies*, www.bibleandscience.com/archaeology/ark.htm, accessed June 20, 2007.

"Who Is the Red Baron?" *essortment*, http://riri.essortment.com/whoredbaron _rkzx.htm, accessed May 23, 2007.

"Who Was Sidney Bechet?" *The Sidney Bechet Society Ltd.* www.sidneybechet.org/ bio.html, accessed May 29, 2007.

"William Grant (General)." *Wikipedia*. http://en.wikipedia.org/wiki/William _Grant_%28General%29, accessed June 2, 2007.

"Woes of Venice Tackled in USA." CNN, 1999, Environmental News Network.

"Wyatt Earp." *San Diego Historical Society*. www.sandiegohistory.org/bio/earp/ earp.htm, accessed June 27, 2007.

"The Young Indiana Jones Chronicles." *The Raider Net*. www.theraider.net/ films/young_indy/, accessed May 25, 2007.

"The Young Indiana Jones Chronicles." *Wikipedia*. http://en.wikipedia.org/wiki/ The_Young_Indiana_Jones_Chronicles, accessed May 25, 2007.

"Young Indiana Jones Videos Episode Guide." *Young Indiana Jones Unofficial Web Site*. www.innermind.com/youngindy/info/yijepg2.htm, accessed May 25, 2007.

Index